ARRL's

EXTRA

Upgrade to an Extra Class
Ham Radio License!

NEW! Fourth Edition

By Ward Silver, NØAX

Contributing Editor:
Mark Wilson, K1RO

Left cover photo: Frandy Johnson, N1FJ
Right cover photo: Sean Kutzko, KX9X

Production Staff:
Jodi Morin, KA1JPA, Assistant Production Supervisor, Layout
Michelle Bloom, WB1ENT, Production Super
Maty Weinberg, KB1EIB, Editorial Assistant
David Pingree, N1NAS, Senior Technical Illus
Sue Fagan, KB1OKW, Graphic Design Super

D1377761

ARRL The national association for
AMATEUR RADIO®
225 Main Street, Newington, CT 06111-1494
www.arrl.org

This book may be used for Amateur Extra license exams
given beginning July 1, 2016. *QST* and the ARRL website
(**www.arrl.org**) will have news about any rules changes affecting
the Extra class license or any of the material in this book.

We strive to produce books without errors. Sometimes mistakes
do occur, however. When we become aware of problems in
our books (other than obvious typographical errors), we post
corrections on the ARRL website. If you think you have found an
error, please check **www.arrl.org/extra-class-license-manual**
for corrections. If you don't find a correction there, please let us
know, either using the Feedback Form at the back of this book or
by sending e-mail to **pubsfdbk@arrl.org**.

Contents

Foreword

Now that you have decided to upgrade to the Extra class Amateur Radio license or are thinking about giving it a try, this book is designed to help. Welcome! Upgrading to Extra will complete your journey to the highest class of license Amateur Radio has to offer.

For more than 100 years, the ARRL has helped radio amateurs get the most out of their hobby by upgrading their licenses and advancing their operating and technical skills. The ARRL license preparation materials are the most complete package available to the amateur. *The ARRL Extra Class License Manual* is a detailed text reference that helps you understand the electronics theory, operating practices and FCC rules. This not only helps you pass the exam but makes you more confident on the air and in your "shack."

All of the information you need to pass the Element 4 Extra class written exam is here in the third edition of *ARRL's Extra Q&A*. Each and every one of the questions in the Extra class examination question pool is addressed with study material that explains the correct answer and provides background information. If you can answer the questions in this book, you can pass the written exam with confidence.

This book and all of the ARRL technical material and operating aids are designed to help you do much more than just pass the written exam. The ARRL's "Radio Amateurs Library" supports almost any amateur operating practice today. The complete publications catalog is online at the ARRL's website — **www.arrl.org** — look for the ARRL Store. While you're there, browse the latest news about Amateur Radio and tap into the wealth of services provided to amateurs by the ARRL. You can also contact the Publication Sales Office at ARRL Headquarters to request the latest publications catalog or to place and order. (You can reach us by phone — 860-594-0200; by fax — 860-594-0303 and by email — **pubsales@arrl.org**.

This fourth edition of *ARRL's Extra Q&A* is the product of cooperation between readers, license class instructors, and the many ARRL staff members. You can help make this book better by providing your own feedback. After you have passed your exam, write your suggestions, questions, and comments on the Feedback Form at the back of the book, and send the form to us. Comments from readers are very important in making subsequent editions more effective and useful to readers.

Thanks for making the decision to upgrade — we hope to hear you on the air soon, using your new Extra class privileges and enjoying more of Amateur Radio. Good luck!

David Sumner, K1ZZ
Chief Executive Officer
Newington, Connecticut
March 2016

New Ham Desk
ARRL Headquarters
225 Main Street
Newington, CT
06111-1494
(860) 594-0200

Prospective new amateurs call:
800-32-NEW-HAM (800-326-3942)
You can also contact us via e-mail:
newham@arrl.org
or check out **ARRLWeb**:
www.arrl.org

What is
Amateur Radio?

Perhaps you've just picked up this book in the library or from a bookstore shelf and are wondering what this Amateur Radio business is all about. Maybe you have a friend or relative who is a "ham" and you're interested in becoming one, as well. In that case, a short explanation is in order.

Amateur Radio or "ham radio" is one of the longest-lived wireless activities. Amateur experimenters were operating right along with Marconi in the early part of the 20th century. They have helped advance the state-of-the-art in radio, television and dozens of other communications services since then, right up to the present day. There are more than 700,000 Amateur Radio operators or "hams" in the United States alone and several million more around the world!

Amateur Radio in the United States is a formal *communications service*, administered by the Federal Communications Commission or FCC. Created officially in its present form in 1934, the Amateur Service is intended to foster electronics and radio experimentation, provide emergency backup communications, encourage private citizens to train and practice operating, and even spread the goodwill of person-to-person contact over the airwaves.

Who Is a Ham and What Do Hams Do?

Anyone can be a ham — there are no age limits or physical requirements that prevent anyone from passing their license exam and getting on the air. Kids as young as 6 years old have passed the basic exam and there are hams out there over the age of 100. You probably fall somewhere in the middle of that range.

Once you get on the air and start meeting other hams, you'll find a wide range of capabilities and interests. Of course, there are many technically skilled hams who work as engineers, scientists or technicians. But just as many don't have a deep technical background. You're just as likely to encounter writers, public safety personnel, students, farmers, truck drivers — anyone with an interest in personal communications over the radio.

The activities of Amateur Radio are incredibly varied. Amateurs who hold the Technician Class license — the usual first license for hams in the US — communicate primarily with local and regional amateurs using relay stations called *repeaters*. Known as "Techs," they sharpen their skills of operating while portable and mobile, often joining emergency communications teams. They may instead focus on the burgeoning wireless data networks assembled and used by hams around the world. Techs can make use of the growing number of Amateur Radio satellites, built and launched by hams along with

Taking ham radio on a vacation trip can be a lot of fun! During a recent trip to Puerto Rico, Sean KX9X made contacts through an amateur satellite from a beach while using a handheld radio!

John, W1RT operates this multiband "rover" during VHF+ contests, driving to hilltops around New England such as from Mohawk Mountain where he is shown giving the antennas a little "hands on" adjustment.

the commercial "birds." Technicians transmit their own television signals, push the limits of signal propagation through the atmosphere and experiment with microwaves. Hams hold most of the world records for long-distance communication on microwave frequencies, in fact!

Arie, PA3A, was one of four Dutch hams who participated in a Mercy Ships project in Sierra Leone. During their free time, they operated as 9L5MS on the HF bands.

Sisters Autumn and Hannah operate K5LBJ during the bi-annual School Club Roundup competition that features the radio clubs at schools across the country.

Hams who advance or *upgrade* to General and then to Extra class are granted additional privileges with each step to use the frequencies usually associated with shortwave operation. This is the traditional Amateur Radio you probably encountered in movies or books. On these frequencies, signals can travel worldwide and so amateurs can make direct contact with foreign hams. No Internet, phone systems, or data networks are required. It's just you, your radio, and the ionosphere — the upper layers of the Earth's atmosphere!

Many hams use voice, Morse code, computer data modes and even image transmissions to communicate. All of these signals are mixed together where hams operate, making the experience of tuning a radio receiver through the crowded bands an interesting experience.

One thing common to all hams is that all of their operation is noncommercial, especially the volunteers who provide emergency communications. Hams pursue their hobby purely for personal enjoyment and to advance their skills, taking satisfaction from providing valuable services to their fellow citizens. This is especially valuable after natural disasters such as hurricanes and earthquakes when commercial systems are knocked out for a while. Amateur operators rush in to provide backup communication for hours, days, weeks or even months until the regular systems are restored. All this from a little study and a simple exam!

Lots of hams enjoy using digital modes to communicate. Melanie, KD0LRC is shown using PSK31 during the KO0A Field Day operation in St Charles, MO.

Want to Find Out More?

If you'd like to find out more about Amateur Radio in general, there is lots of information available on the Internet. A good place to start is on the American Radio Relay League's (ARRL) ham radio introduction page at **www.hello-radio.org**. Books like *Ham Radio for Dummies* and *Getting Started*

Participating in a "radiosport" competition is a great way to build up your radio skills. The W2GD team specializes in 160 meter operation as shown here during the ARRL 160 Meter Contest.

With Ham Radio will help you "fill in the blanks" as you learn more.

Along with books and Internet pages, there is no better way to learn about ham radio than to meet your local amateur operators. It is quite likely that no matter where you live in the United States, there is a ham radio club in your area — perhaps several! The ARRL provides a club lookup web page at **www.arrl.org** where you can find a club just by entering your Zip code or state. Carrying on the tradition of mutual assistance, many clubs make helping newcomers to ham radio a part of their charter.

If this sounds like hams are confident that you'll find their activities interesting, you're right! Amateur Radio is much more than just talking on a radio, as you'll find out. It's an opportunity to dive into the fascinating world of radio communications, electronics, and computers as deeply as you wish to go. Welcome!

When to Expect New Books

A Question Pool Committee (QPC) consisting of representatives from the various Volunteer Examiner Coordinators (VECs) prepares the license question pools. The QPC establishes a schedule for revising and implementing new question pools. The current question pool revision schedule is as follows:

Question Pool	Current Study Guides	Valid Through
Technician (Element 2)	*The ARRL Ham Radio License Manual,* 3rd Edition *ARRL's Tech Q&A,* 6th Edition	June 30, 2018
General (Element 3)	*The ARRL General Class License Manual,* 8th Edition *ARRL's General Q&A,* 5th Edition	June 30, 2019
Amateur Extra (Element 4)	*The ARRL Extra Class License Manual,* 11th Edition *ARRL's Extra Q&A,* 4th Edition	June 30, 2020

As new question pools are released, ARRL will produce new study materials before the effective date of the new pools. Until then, the current question pools will remain in use, and current ARRL study materials, including this book, will help you prepare for your exam.

As the new question pool schedules are confirmed, the information will be published in *QST* and on the ARRL website at **www.arrl.org**.

How to Use
This Book

To earn an Extra class Amateur Radio license, you must pass (or receive credit for) FCC Elements 2 (Technician class), 3 (General class) and 4 (Extra class). This book is designed to help you prepare for and pass the Element 4 written exam. If you do not already have a General class Amateur Radio license, you will need some additional study materials for the Element 3 (General) exam.

The Element 4 exam consists of 50 questions about Amateur Radio rules, theory and practice, as well as some electronics and radio theory. A passing grade is 74%, so you must answer 37 of the 50 questions correctly.

ARRL's Extra Q&A has 10 major sections that follow subelements E1 through E0 in the Extra class syllabus. The questions and multiple choice answers in this book are printed exactly as they were written by the Volunteer Examiner Coordinators' Question Pool Committee, and exactly as they will appear on your exam. (Be careful, though. The letter position of the answer choices may be scrambled, so you can't simply memorize an answer letter for each question.) In this book, the letter of the correct answer is printed in **boldface** type just before the explanation.

The ARRL also maintains a special web page for Extra Class students at **www.arrl.org/extra-class-license-manual**. The useful and interesting on-line references listed there put you one click away from related and useful information.

If you are taking a licensing class, help your instructors by letting them know about areas in which you need help. They want you to learn as thoroughly and quickly as possible, so don't hold back with your questions. Similarly, if you find the material particularly clear or helpful, tell them that, too, so it can be used in the next class!

What We Assume About You

You don't have to be a technical guru or an expert operator to upgrade to Extra class! As you progress through the material, you'll build on the radio and electronics concepts you mastered for previous license exams. No advanced mathematics is needed and if math gives you trouble, tutorials are listed at **www.arrl.org/extra-class-license-manual**. As with the General license, mastering rules and regulations will require learning some definitions and the finer points of regulations you may have already encountered. You should have a simple calculator, which you'll also be allowed to use during the license exam.

If you have some background in radio, perhaps as a technician or trained operator, you may be able to short-circuit some of the sections. It's common for technically-minded students to focus on the rules and regulations while students with an operating background tend to need the technical material more. Whichever you may be, be sure that you can answer the questions because they will certainly be on the test!

ARRL's Extra Q&A can be used either by an individual student, studying on his or her own, or as part of a licensing class taught by an instructor. If you're part of a class, the instructor will set the order in which the material is covered. The solo student can move at any pace and in any convenient order. You'll find that having a buddy to study with makes learning the material more fun as you help each other over the rough spots.

Don't hesitate to ask for help! If you can't find the answer in the book or at the website, email your question to the ARRL's New Ham Desk, **newham@arrl. org**. The ARRL's experts will answer directly or connect you with another ham that can answer your questions.

Online Exams

While you're studying and when you feel like you're ready for the actual exam you can get some good practice by taking one of the on-line Amateur Radio exams. These websites use the same Question Pool to construct an exam with the same number and variety of questions that you'll encounter on exam day. The exams are free and you can take them over and over again in complete privacy. Links to on-line exams can be found at **www.arrl.org/exam-practice**.

These exams are quite realistic and you get quick feedback about the questions you missed. When you find yourself passing the on-line exams by a comfortable margin, you'll be ready for the real thing! A note of caution, be sure that the questions used are from the current question pool.

About
The ARRL

The seed for Amateur Radio was planted in the 1890s, when Guglielmo Marconi began his experiments in wireless telegraphy. Soon he was joined by dozens, then hundreds, of others who were enthusiastic about sending and receiving messages through the air — some with a commercial interest, but others solely out of a love for this new communications medium. The United States government began licensing Amateur Radio operators in 1912.

By 1914, there were thousands of Amateur Radio operators — hams — in the United States. Hiram Percy Maxim, a leading Hartford, Connecticut inventor and industrialist, saw the need for an organization to unify this fledgling group of radio experimenters. In May 1914 he founded the American Radio Relay League (ARRL) to meet that need.

ARRL is the national association for Amateur Radio in the US. Today, with approximately 170,000 members, ARRL numbers within its ranks the vast majority of active radio amateurs in the nation and has a proud history of achievement as the standard-bearer in amateur affairs. ARRL's underpinnings as Amateur Radio's witness, partner, and forum are defined by five pillars: Public Service, Advocacy, Education, Technology, and Membership. ARRL is also International Secretariat for the International Amateur Radio Union, which is made up of similar societies in 150 countries around the world.

ARRL's Mission Statement: To advance the art, science, and enjoyment of Amateur Radio.

ARRL's Vision Statement: As the national association for Amateur Radio in the United States, ARRL:

- Supports the awareness and growth of Amateur Radio worldwide;
- Advocates for meaningful access to radio spectrum;
- Strives for every member to get involved, get active, and get on the air;
- Encourages radio experimentation and, through its members, advances radio technology and education; and
- Organizes and trains volunteers to serve their communities by providing public service and emergency communications.

At ARRL headquarters in the Hartford, Connecticut suburb of Newington, the staff helps serve the needs of members. ARRL publishes the monthly journal *QST* and an interactive digital version of *QST*, as well as newsletters and many publications covering all aspects of Amateur Radio. Its headquarters station, W1AW, transmits bulletins of interest to radio amateurs and Morse code practice sessions. ARRL also coordinates an extensive field organization, which includes volunteers who provide technical information and other support services for radio amateurs as well as communications for public service activities. In addition, ARRL represents US radio amateurs to the Federal Communications Commission and other government agencies in the US and abroad.

Membership in ARRL means much more than receiving *QST* each month. In addition to the services already described, ARRL offers membership services on a personal level, such as the Technical Information Service, where members can get answers — by phone, e-mail, or the ARRL website — to all their technical and operating questions.

A bona fide interest in Amateur Radio is the only essential qualification of membership; an Amateur Radio license is not a prerequisite, although full voting membership is granted only to licensed radio amateurs in the US. Full ARRL membership gives you a voice in how the affairs of the organization are governed. ARRL policy is set by a Board of Directors (one from each of 15 Divisions). Each year, one-third of the ARRL Board of Directors stands for election by the full members they represent. The day-to-day operation of ARRL HQ is managed by a Chief Executive Officer and his/her staff.

Join ARRL Today! No matter what aspect of Amateur Radio attracts you, ARRL membership is relevant and important. There would be no Amateur Radio as we know it today were it not for ARRL. We would be happy to welcome you as a member! Join online at **www.arrl.org/join**. For more information about ARRL and answers to any questions you may have about Amateur Radio, write or call:

ARRL — The national association for Amateur Radio®
225 Main Street
Newington CT 06111-1494
Tel: 860-594-0200
FAX: 860-594-0259
e-mail: **hq@arrl.org**
www.arrl.org

Prospective new radio amateurs call (toll-free):
800-32-NEW HAM (800-326-3942)
You can also contact ARRL via e-mail at **newham@arrl.org**
or check out the ARRL website at **www.arrl.org**

Books to Help
You Learn

As you study the material on the licensing exam, you will have lots of other questions about the hows and whys of Amateur Radio. The following references, available from your local dealer or the ARRL (**www.arrl.org/shop**) will help "fill in the blanks" and give you a broader picture of the hobby:

• *Ham Radio for Dummies, Second Edition*, by Ward Silver, NØAX. Written for new hams and hams interested in new activities, this book supplements the information in study guides with an informal, friendly approach to the hobby.

• *ARRL Operating Manual*. Devoted to on-the-air activities, there are in-depth chapters on award programs, DXing, contests, digital and image modes and more, with a healthy set of reference tables and maps.

• *Understanding Basic Electronics* by Walter Banzhaf, WB1ANE. Students who want who want more technical background about radio and electronics should take a look at this book. The book covers the fundamentals of electricity and electronics that are the foundation of all radio.

• *Basic Radio* by Joel Hallas, W1ZR. This book goes beyond electronic circuits to explain how radios are designed and perform. It covers the key building block of receivers, transmitters, antennas and propagation.

• *ARRL Handbook*. This is the grandfather of all Amateur Radio references and belongs on the shelf of all hams. Almost any topic you can think of in Amateur Radio technology is represented here.

• *ARRL Antenna Book*. After the radio itself, all radio depends on antennas. This book provides information on every common type of amateur antenna, feed lines and related topics, and construction tips and techniques.

Introduction

Welcome to *ARRL's Extra Q&A*! Obtaining your Extra class license will enable you to make the most of your Amateur Radio experience. The additional HF band segments reserved for Extra class operators are the best for DXing and contesting, for example. Studying for the additional technical topics covered by the exam will help you get the most out of your current operating preferences.

This study guide will not only teach you the answers to the Extra class exam questions, but also provides explanations and supporting information. That way, you'll find it easier to learn the basic principles involved, and that will help you remember what you've learned. The book is full of useful facts and figures, so you'll want to keep it handy after you pass the test and are using your new privileges.

Do You Have a General License?

Most of this book's readers will have already earned their General class license, and some will have an Advanced license. Some readers may have been a ham for quite a while, and others may be new to the hobby. In either case, you're to be commended for making the effort to upgrade. We'll try to make it easy to pass your exam by teaching you the fundamentals and rationale behind each question and answer.

Reasons to Upgrade

If you're browsing through this book, trying to decide whether or not to upgrade, here are a few good reasons:

• *More fun.* The Extra class licensee has access to reserved sections of several of the HF bands. These *Extra class segments* are where most DX contacts are made and are considered prime territory. See **Table 1** for details of all of the frequencies reserved exclusively for Extra class licensees.

• *More communications options.* Your new understanding and skills will be valuable to your club, operating team or community.

• *New technical opportunities.* With your new understanding will come new ways of assembling and operating a station. Your improved technical understanding of electronics, radio and propagation will make you a more knowledgeable and skilled operator.

• *Volunteer examinations.* As an Extra class licensee, you'll be able to administer exams for any license class. Amateur Radio needs your help to make the volunteer licensing program work.

Not only does upgrading grant you complete Amateur Radio privileges, but by learning the material — perhaps even learning to operate using Morse code — your experiences will be much broader. You'll enjoy the hobby in ways that hams have pioneered and fostered for generations. The Extra's privileges are well worth your effort!

Extra Class Overview

There are three classes of license being granted today: Technician, General and Amateur Extra. Each grants the licensee more and more privileges, meaning access to frequencies and modes. Extra class licensees have all amateur privileges. **Table 2** shows the elements for each of the amateur licenses as of early 2016.

Table 1
Extra Class Band Segments

Band	Frequencies (MHz)
80 meters	3.500 – 3.525 and 3.600 – 3.700
40 meters	7.000 – 7.025
20 meters	14.000 – 14.025 and 14.150 – 14.175
15 meters	21.000 – 21.025 and 21.200 – 21.225

Table 2
Amateur License Class Examinations

License Class	Elements Required	Number of Questions
Technician	2 (Written)	35 (passing is 26 correct)
General	3 (Written)	35 (passing is 26 correct)
Extra	4 (Written)	50 (passing is 37 correct)

Table 3
Exam Elements Needed to Qualify for an Extra Class License

Current License	Exam Requirements	Study Materials
None or Novice	Technician (Element 2)	The ARRL Ham Radio License Manual or ARRL's Tech Q&A
	General (Element 3)	The ARRL General Class License Manual or ARRL's General Q&A
	Amateur Extra (Element 4)	The ARRL Extra Class License Manual or ARRL's Extra Q&A
Technician (issued on or after March 21, 1987)*	General (Element 3)	The ARRL General Class License Manual or ARRL's General Q&A
	Amateur Extra (Element 4)	The ARRL Extra Class License Manual or ARRL's Extra Q&A
General or Advanced	Amateur Extra (Element 4)	The ARRL Extra Class License Manual or ARRL's Extra Q&A

*Individuals who qualified for the Technician license before March 21, 1987, will be able to receive credit for Element 3 (General class) by providing documentary proof to a Volunteer Examiner Coordinator.

As shown in Table 2, to qualify for an Extra class license, you must have passed Elements 2 (Technician), 3 (General) and 4 (Extra). If you hold a General or Advanced license, you are credited with Elements 2 and 3, so you don't have to take them again. If you have a Technician license issued before March 21, 1987, you can upgrade to General simply by going to a test session with proof of being licensed before that date. **Table 3** shows the variations of licenses and element credit.

The 50-question multiple-choice test for Element 4 is more comprehensive

Stu Turner, WØSTU operated from the top of Colorado's Mount Herman in the ARRL January VHF Sweepstakes. Operating on 50, 144, and 432 MHz , he earned a spot in the Single-Operator, Portable categories Top Ten!

than the Element 3 General exam because as a more experienced ham, your wider knowledge will allow you to experiment with, modify, and build more types of equipment and antennas. An Extra class amateur can use any mode, frequency or technique available to amateurs.

Morse Code

Although you no longer need to learn Morse code for any license exam, Morse code or "CW" has part of the rich amateur tradition for 100 years and many hams still use it extensively. There are solid reasons for it to be used, too! It's easy to build Morse code transmitters and receivers. There is no more power-efficient mode of communication that is copied by the human ear. The extensive set of prosigns and signals allows amateurs to communicate a great deal of information even if they don't share a common language. Morse is likely to remain part of the amateur experience for a long time to come.

If you are interested in learning Morse code, the ARRL has a complete set of resources listed on its web page at **www.arrl.org/learning-morse-code**. Computer software and on-the-air code practice sessions are available for personal training and practice. Organizations such as FISTS (**www.fists.org**) — an operator's style of sending is referred to as his or her "fist" — help hams learn Morse code and will even help you find a "code buddy" to share the learning with you. CWops (**www.cwops.org**) offers a CW Academy to help amateurs learn Morse code or improve their CW skills.

Want More Information?

Looking for more information about Extra class instruction in your area? Are you ready to take the Extra class exam? Do you need a list of ham radio clubs, instructors or examiners in your local area? The following web pages are very helpful in finding the local resources you need to successfully pass your Extra exam:

www.arrl.org — the ARRL's home page, it features news and links to other ARRL resources

www.arrl.org/extra-class-license-manual — the website that supports this book

www.arrl.org/find-a-club— a search page to find ARRL-Affiliated clubs

www.arrl.org/exam — the ARRL VEC exam session search page

www.arrl.org/technical-information-service — the ARRL's Technical Information Service is an excellent resource

The Upgrade Trail

As you begin your studies remember that you've already overcome two big hurdles — taking and passing the Technician and General license exams! The questions may be more numerous and challenging for the Extra class exam, but you already know all about the testing procedure and the basics of ham radio. You can approach the process of upgrading with confidence!

Refine Your Knowledge

The Extra class exam mostly deals with the finer points of operating, the FCC's rules, and the technical details of radio. We'll cover more advanced modes and signals, too. The goal is to help you "fill in the blanks" in your ham radio knowledge. Here are some examples of topics that you'll be studying:

- Impedance and resonance
- Image modes — ATV and SSTV
- How digital protocols work
- Long-path, grey line and transequatorial propagation
- Receiver performance — for both analog and software-defined radios
- Practical radio circuits
- More types of antennas

Not every ham uses every mode and frequency, of course. By mastering a wider range of skills and knowledge, hams can get more out of everyday operating. Extra class hams can take the lead in planning and assembling stations and operating teams. Having deeper knowledge of radio will lead you to a greater appreciation of the magic of radio!

Testing Process

When you're ready, you'll need to find a test session. If you're in a licensing class, the instructor will help you find and register for a session. Otherwise, you can find a test session by using the ARRL's web page for finding exams, **www. arrl.org/find-an-amateur-radio-license-exam-session**. If you can register for the test session in advance, do so. Other sessions, such as those at hamfests or conventions, are available to anyone that shows up (called *walk-ins*). You may have to wait for an available space though, so go early!

As for all amateur exams, the Extra class exam is administered by Volunteer Examiners (VEs). All VEs are certified by a Volunteer Examiner Coordinator (VEC) such as the ARRL VEC. This organization trains and certifies VEs and processes the FCC paperwork for their test sessions.

Bring the *original* of your current license and a photocopy (to send with the application). You'll need two forms of identification including at least one photo ID, such as a driver's license, passport or employer's identity card. Know your

Social Security Number (SSN). You can bring pencils or pens, blank scratch paper and a calculator, but any kind of computer or on-line device is prohibited.

The FCC allows Volunteer Examiners to use a range of procedures to accommodate applicants with various disabilities. If this applies to you, you'll still have to pass the test, but special exam procedures can be applied. Contact your local VE team or the Volunteer Examiner Coordinator (VEC) responsible for the test session you'll be attending. Contact the ARRL VEC Office at 225 Main St, Newington CT 06111-1494 or by phone at 860-594-0200.

Once you're signed in, you'll need to fill out a copy of the National Conference of Volunteer Examiner Coordinator's (NCVEC) Quick Form 605. This is an application for a new or upgraded license. It is used only at test sessions and for a VEC to process a license renewal or a license change. *Do not* use an

Kenneth Ransom, N5VHO is the ISS Ham Radio Project Coordinator and is shown here demonstrating the VHF/UHF transceiver installed in the ISS used to contact Earth-bound hams.

NCVEC Quick Form 605 for any kind of application directly to the FCC — it will be rejected. Use a regular FCC Form 605. After filling out the form, pay the current test fee and get ready.

The Exam

To complete the Extra exam takes less than an hour. You will be given a question booklet and an answer sheet. Be sure to read the instructions, fill in all the necessary information and sign your name wherever required. Check to be sure your booklet has all the questions and be sure to mark the answer in the correct space for each question.

You don't have to answer the questions in order — skip the hard ones and go back to them. If you read the answers carefully, you'll probably find that you can eliminate one or more "distracters." Of the remaining answers, only one will be the best. If you can't decide which is the correct answer, go ahead and guess. There is no penalty for an incorrect guess. When you're done, go back and check your answers and double-check your arithmetic — there's no rush!

Once you've answered all 50 questions, the VEs will grade and verify your test results. Assuming you've passed (congratulations!) you'll fill out a *Certificate of Successful Completion of Examination* (CSCE). The exam organizers will submit your results to the FCC while you keep the CSCE as evidence that you've passed your Amateur Extra test.

NCVEC QUICK-FORM 605 APPLICATION FOR AMATEUR OPERATOR/PRIMARY STATION LICENSE

SECTION 1 - TO BE COMPLETED BY APPLICANT

PRINT LAST NAME: MORIN   SUFFIX (Jr., Sr.)   FIRST NAME: Joanne   INITIAL: B   STATION CALL SIGN (IF ANY): KA1JPA

MAILING ADDRESS (Number and Street or P.O. Box): 225 Main St.   SOCIAL SECURITY NUMBER (SSN) or (FRN) FCC FEDERAL REGISTRATION NUMBER: 987-654-321

CITY: Newington   STATE CODE: CT   ZIP CODE (5 or 9 Numbers): 06067   E-MAIL ADDRESS (OPTIONAL):

DAYTIME TELEPHONE NUMBER (Include Area Code) OPTIONAL: FAX NUMBER (Include Area Code) OPTIONAL: ENTITY NAME (IF CLUB, MILITARY RECREATION, RACES):

Type of Applicant: ☒ Individual   ☐ Amateur Club   ☐ Military Recreation   ☐ RACES (Modify Only)   CLUB, MILITARY RECREATION, OR RACES CALL SIGN:

I HEREBY APPLY FOR (Make an X in the appropriate box(es))   SIGNATURE OF RESPONSIBLE CLUB OFFICIAL (not trustee)

☐ EXAMINATION for a new license grant
☒ EXAMINATION for upgrade of my license class
☐ CHANGE my name on my license to my new name
☐ CHANGE my mailing address to above address
☐ CHANGE my station call sign systematically   Applicant's Initials: _____
☐ RENEWAL of my license grant.

Former Name: _____
(Last name) (Suffix)   (First name)   (MI)

Do you have another license application on file with the FCC which has not been acted upon?   PURPOSE OF OTHER APPLICATION   PENDING FILE NUMBER (FOR VEC USE ONLY)

I certify that:
- I waive any claim to the use of any particular frequency regardless of prior use by license or otherwise;
- All statements and attachments are true, complete and correct to the best of my knowledge and belief and are made in good faith;
- I am not a representative of a foreign government;
- I am not subject to a denial of Federal benefits pursuant to Section 5301of the Anti-Drug Abuse Act of 1988, 21 U.S.C. § 862;.
- The construction of my station will NOT be an action which is likely to have a significant environmental effect (See 47 CFR Sections 1.1301-1.1319 and Section 97.13(a));
- I have read and WILL COMPLY with Section 97.13(c) of the Commission's Rules regarding RADIOFREQUENCY (RF) RADIATION SAFETY and the amateur service section of OST/OET Bulletin Number 65.

Signature of applicant (Do not print, type, or stamp. Must match applicant's name above.) (Clubs: 2 different individuals must sign)

X _Joanne B Morin_   Date Signed: 2/23/16

SECTION 2 - TO BE COMPLETED BY ALL ADMINISTERING VEs

Applicant is qualified for operator license class:

☐ NO NEW LICENSE OR UPGRADE WAS EARNED
☐ TECHNICIAN   Element 2
☐ GENERAL   Elements 2 and 3
☒ AMATEUR EXTRA   Elements 2, 3 and 4

DATE OF EXAMINATION SESSION:
EXAMINATION SESSION LOCATION:
VEC ORGANIZATION:
VEC RECEIPT DATE:

I CERTIFY THAT I HAVE COMPLIED WITH THE ADMINISTERING VE REQUIRMENTS IN PART 97 OF THE COMMISSION'S RULES AND WITH THE INSTRUCTIONS PROVIDED BY THE COORDINATING VEC AND THE FCC.

	VEs STATION CALL SIGN	VEs SIGNATURE (Must match name)	DATE SIGNED
1st VEs NAME (Print First, MI, Last, Suffix): Steven R. Ewald	WV1X	Steven R Ewald	2/23-16
2nd VEs NAME (Print First, MI, Last, Suffix): Rose-Anne Lawrence	KB1DMW	Rose-Anne Lawrence	2-23-16
3rd VEs NAME (Print First, MI, Last, Suffix): Penny E Harts	N1NAC	Penny E Harts	2-23-16

DO NOT SEND THIS FORM TO FCC – THIS IS NOT AN FCC FORM.
IF THIS FORM IS SENT TO FCC, FCC WILL RETURN IT TO YOU WITHOUT ACTION.

NCVEC FORM 605 - February 2007
FOR VE/VEC USE ONLY - Page 1

ARRL0138

This sample NCVEC Quick Form 605 shows how your form will look after you have completed your upgrade to Extra.

If you are licensed and already have a call sign, you can begin using your new privileges immediately. When you give your call sign, append "/AE" (on CW or digital modes) or "temporary AE" (on phone). As soon as your name and call sign appear in the FCC's database of licensees, typically a week to 10 days later, you can stop adding the suffix. The CSCE is good for 365 days in case you need to re-take an exam before receiving your paper license.

If you don't pass, don't be discouraged! You might be able to take another version of the test right then and there if the session organizers can accommodate you. Even if you decide to try again later, you now know just how the test session feels — you'll be more relaxed and ready next time. The ham bands are full of hams who took their Extra test more than once before passing. You'll be in good company!

FCC and ARRL VEC Licensing Resources

After you pass your exam, the examiners will file all of the necessary paperwork so that your license will be granted by the Federal Communications Commission (FCC). In a week or two, you will be able see your new license status — and your new call sign if you requested one — in the FCC's database via the ARRL website. The ARRL VEC can also process license renewals and modifications for you as described at **www.arrl.org/call-sign-renewals-or-changes**.

When you initially passed your Technician exam, you may have applied for your FCC Federal Registration Number (FRN) or may have been issued one automatically by FCC from the information gathered from your form. This allows you to access the information for any FCC licenses you may have and to request modifications to them. These functions are available via the FCC's Universal Licensing System (ULS) website (**wireless.fcc.gov/uls**) and complete instructions for using the site are available at **www.arrl.org/universal-licensing-system**. When accessing the ULS, your FRN will allow you to watch the database for your license upgrade!

Time to Get Started

By following these instructions and carefully studying the material in this book, soon you'll be operating in the Extra class segments with the rest of the Extras! Each of us at the ARRL Headquarters and every ARRL member looks forward to the day when you join the fun. '73' (best regards) and good luck!

Extra Class (Element 4) Syllabus

Valid July 1, 2016 through June 30, 2020

SUBELEMENT E1 — COMMISSION'S RULES
[6 Exam Questions — 6 Groups]

E1A — Operating standards: frequency privileges; emission standards; automatic message forwarding; frequency sharing; stations aboard ships or aircraft.

E1B — Station restrictions and special operations: restrictions on station location; general operating restrictions, spurious emissions, control operator reimbursement; antenna structure restrictions; RACES operations; national quiet zone.

E1C — Definitions and restrictions pertaining to local, automatic and remote control operation; control operator responsibilities for remote and automatically controlled stations; IARP and CEPT licenses; third party communications over automatically controlled stations.

E1D — Amateur satellites: definitions and purpose; license requirements for space stations; available frequencies and bands; telecommand and telemetry operations; restrictions, and special provisions; notification requirements.

E1E — Volunteer examiner program: definitions; qualifications; preparation and administration of exams; accreditation; question pools; documentation requirements.

E1F — Miscellaneous rules: external RF power amplifiers; business communications; compensated communications; spread spectrum; auxiliary stations; reciprocal operating privileges; special temporary authority.

SUBELEMENT E2 — OPERATING PROCEDURES
[5 Exam Questions — 5 Groups]

E2A — Amateur radio in space: amateur satellites; orbital mechanics; frequencies and modes; satellite hardware; satellite operations; experimental telemetry applications.

E2B — Television practices: fast scan television standards and techniques; slow scan television standards and techniques.

E2C — Operating methods: contest and DX operating; remote operation techniques; Cabrillo format; QSLing; RF network connected systems.

E2D — Operating methods: VHF and UHF digital modes and procedures; APRS; EME procedures, meteor scatter procedures.

E2E — Operating methods: operating HF digital modes

SUBELEMENT E3 — RADIO WAVE PROPAGATION

[3 Exam Questions — 3 Groups]

E3A — Electromagnetic waves; Earth-Moon-Earth communications; meteor scatter; microwave tropospheric and scatter propagation; aurora propagation.

E3B — Transequatorial propagation; long path; gray-line; multi-path; ordinary and extraordinary waves; chordal hop, sporadic E mechanisms.

E3C — Radio-path horizon; less common propagation modes; propagation prediction techniques and modeling; space weather parameters and amateur radio.

SUBELEMENT E4 — AMATEUR PRACTICES

[5 Exam Questions — 5 Groups]

E4A — Test equipment: analog and digital instruments; spectrum and network analyzers, antenna analyzers; oscilloscopes; RF measurements; computer aided measurements.

E4B — Measurement technique and limitations: instrument accuracy and performance limitations; probes; techniques to minimize errors; measurement of "Q"; instrument calibration; S parameters; vector network analyzers.

E4C — Receiver performance characteristics, phase noise, noise floor, image rejection, MDS, signal-to-noise-ratio; selectivity; effects of SDR receiver non-linearity.

E4D — Receiver performance characteristics: blocking dynamic range; intermodulation and cross-modulation interference; 3rd order intercept; desensitization; preselector.

E4E — Noise suppression: system noise; electrical appliance noise; line noise; locating noise sources; DSP noise reduction; noise blankers; grounding for signals.

SUBELEMENT E5 — ELECTRICAL PRINCIPLES

[4 Exam Questions — 4 Groups]

E5A — Resonance and Q: characteristics of resonant circuits: series and parallel resonance; definitions and effects of Q; half-power bandwidth; phase relationships in reactive circuits.

E5B — Time constants and phase relationships: RLC time constants; definition; time constants in RL and RC circuits; phase angle between voltage and current; phase angles of series RLC; phase angle of inductance vs susceptance; admittance and susceptance.

E5C — Coordinate systems and phasors in electronics: rectangular coordinates; polar coordinates; phasors.

E5D — AC and RF energy in real circuits: skin effect; electrostatic and electromagnetic fields; reactive power; power factor; electrical length of conductors at UHF and microwave frequencies.

SUBELEMENT E6 — CIRCUIT COMPONENTS

[6 Exam Questions – 6 Groups]

E6A — Semiconductor materials and devices: semiconductor materials; germanium, silicon, P-type, N-type; transistor types: NPN, PNP, junction, field-effect transistors: enhancement mode; depletion mode; MOS; CMOS; N-channel; P-channel.

E6B — Diodes.

E6C — Digital ICs: families of digital ICs; gates; programmable logic devices (PLDs).

E6D — Toroidal and solenoidal inductors: permeability, core material, selecting, winding; transformers; piezoelectric devices.

E6E — Analog ICs: MMICs, CCDs, Device packages.

E6F — Optical components: photoconductive principles and effects, photovoltaic systems, optical couplers, optical sensors, and optoisolators; LCDs.

SUBELEMENT E7 — PRACTICAL CIRCUITS

[8 Exam Questions – 8 Groups]

E7A — Digital circuits: digital circuit principles and logic circuits: classes of logic elements; positive and negative logic; frequency dividers; truth tables.

E7B — Amplifiers: class of operation; vacuum tube and solid-state circuits; distortion and intermodulation; spurious and parasitic suppression; microwave amplifiers; switching-type amplifiers.

E7C — Filters and matching networks: types of networks; types of filters; filter applications; filter characteristics; impedance matching; DSP filtering.

E7D — Power supplies and voltage regulators; solar array charge controllers.

E7E — Modulation and demodulation: reactance, phase and balanced modulators; detectors; mixer stages.

E7F — DSP filtering and other operations; software defined radio fundamentals; DSP modulation and demodulation.

E7G — Active filters and op-amp circuits: active audio filters; characteristics; basic circuit design; operational amplifiers.

E7H — Oscillators and signal sources: types of oscillators; synthesizers and phase-locked loops; direct digital synthesizers; stabilizing thermal drift; microphonics; high accuracy oscillators.

SUBELEMENT E8 — SIGNALS AND EMISSIONS

[4 Exam Questions — 4 Groups]

E8A — AC waveforms: sine, square, sawtooth and irregular waveforms; AC measurements; average and PEP of RF signals; Fourier analysis; analog to digital conversion: digital to analog conversion.

E8B — Modulation and demodulation: modulation methods; modulation index and deviation ratio; frequency and time division multiplexing; orthogonal frequency division multiplexing.

E8C — Digital signals: digital communication modes; information rate vs bandwidth; error correction.

E8D — Keying defects and overmodulation of digital signals; digital codes; spread spectrum.

SUBELEMENT E9 — ANTENNAS AND TRANSMISSION LINES

[8 Exam Questions — 8 Groups]

E9A — Basic antenna parameters: radiation resistance, gain, beamwidth, efficiency, beamwidth; effective radiated power, polarization.

E9B — Antenna patterns: E and H plane patterns; gain as a function of pattern; antenna design.

E9C — Wire and phased array antennas: rhombic antennas; effects of ground reflections; take-off angles; practical wire antennas: Zepps, OCFD, loops.

E9D — Directional antennas: gain; Yagi antennas; losses; SWR bandwidth; antenna efficiency; shortened and mobile antennas; RF grounding.

E9E — Matching: matching antennas to feed lines; phasing lines; power dividers.

E9F — Transmission lines: characteristics of open and shorted feed lines; ⅛ wavelength; ¼ wavelength; ½ wavelength; feed lines: coax versus open-wire; velocity factor; electrical length; coaxial cable dielectrics; velocity factor.

E9G — The Smith chart.

E9H — Receiving antennas: radio direction finding antennas; Beverage antennas; specialized receiving antennas; longwire receiving antennas.

SUBELEMENT E0 — SAFETY

[1 Exam Question — 1 Group]

E0A — Safety: amateur radio safety practices; RF radiation hazards; hazardous materials; grounding and bonding.

Commission's Rules

The Extra class (Element 4) written examination consists of 50 questions taken from the Extra class examination pool. This pool is prepared by the Volunteer Examiner Coordinators' Question Pool Committee. A certain number of questions are taken from each of the 10 subelements (numbered E1 through E0).

There will be six examination questions over the six groups of questions covering the Commission's Rules. The question groups are labeled E1A through E1F.

The correct answer (A, B, C or D) is given in bold following the question and the possible responses at the beginning of an explanation section. This convention will be used throughout this book.

After many of the explanations in this subelement, you will see a reference to Part 97 of the FCC rules set inside square brackets, like [97.301(b)]. This tells you where to look for the exact wording in the Rules as they relate to that question. For a complete copy of Part 97, along with simple explanations of the Rules governing Amateur Radio, see the FCC Rules online at **www.arrl.org/part-97-amateur-radio**. In addition to Part 97, you'll find other parts of the FCC rules. These include Part 1 and Part 17.

E1A Operating standards: frequency privileges; emission standards; automatic message forwarding; frequency sharing; stations aboard ships or aircraft

E1A01 When using a transceiver that displays the carrier frequency of phone signals, which of the following displayed frequencies represents the highest frequency at which a properly adjusted USB emission will be totally within the band?

A. The exact upper band edge
B. 300 Hz below the upper band edge
C. 1 kHz below the upper band edge
D. 3 kHz below the upper band edge

D Figure E1A01 illustrates the relationship between the carrier frequency and the frequency of the sideband components that make up the single-sideband signal. Most transceivers are configured to show the carrier frequency of a signal. The components that make up the sideband of a USB signal are higher than the carrier frequency. Since an amateur SSB signal generally has a bandwidth of about 3 kHz, to be sure the sideband components of a USB signal are within the amateur band, the carrier frequency should be 3 kHz below the band edge. [97.301, 97.305]

E1A02 When using a transceiver that displays the carrier frequency of phone signals, which of the following displayed frequencies represents the lowest frequency at which a properly adjusted LSB emission will be totally within the band?

A. The exact lower band edge
B. 300 Hz above the lower band edge
C. 1 kHz above the lower band edge
D. 3 kHz above the lower band edge

D See E1A01.

E1A03 With your transceiver displaying the carrier frequency of phone signals, you hear a DX station calling CQ on 14.349 MHz USB. Is it legal to return the call using upper sideband on the same frequency?

A. Yes, because you were not the station calling CQ
B. Yes, because the displayed frequency is within the 20 meter band
C. No, the sideband will extend beyond the band edge
D. No, U.S. stations are not permitted to use phone emissions above 14.340 MHz

C See E1A01.

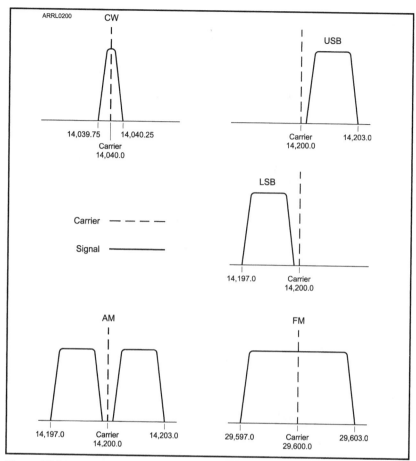

Figure E1A01 — An illustration of the relationship between carrier frequency and the actual signal sidebands. Most transceivers are configured to display the carrier frequency and not the frequency of the sidebands so the amateur needs to take care to keep all of the signal in the band.

E1A04 With your transceiver displaying the carrier frequency of phone signals, you hear a DX station calling CQ on 3.601 MHz LSB. Is it legal to return the call using lower sideband on the same frequency?

A. Yes, because the DX station initiated the contact

B. Yes, because the displayed frequency is within the 75 meter phone band segment

C. No, the sideband will extend beyond the edge of the phone band segment

D. No, U.S. stations are not permitted to use phone emissions below 3.610 MHz

C See E1A01.

US Amateur Bands

Effective Date March 5, 2012

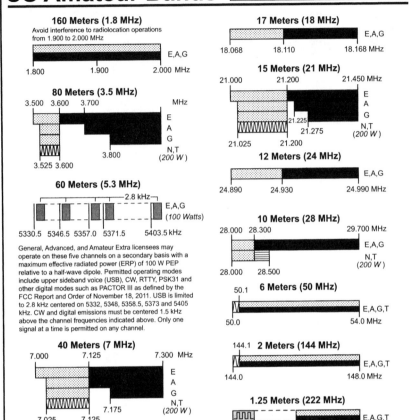

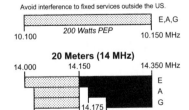

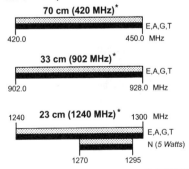

Figure E1A05 — Amateur operating privileges

E1A05 What is the maximum power output permitted on the 60 meter band?

A. 50 watts PEP effective radiated power relative to an isotropic radiator

B. 50 watts PEP effective radiated power relative to a dipole

C. 100 watts PEP effective radiated power relative to the gain of a half-wave dipole

D. 100 watts PEP effective radiated power relative to an isotropic radiator

C If an antenna other than a half-wavelength dipole is used, the transmitter output power must be adjusted to account for the gain of the antenna. For example, if the antenna has a gain of 3 dBd (twice the gain of a dipole), the maximum transmitter output power is 50 watts. If an antenna with less gain than a dipole is used, more than 100 watts can be used. See the material in Subelement E9 for more information on effective radiated power (ERP). [97.313(i)]

E1A06 Where must the carrier frequency of a CW signal be set to comply with FCC rules for 60 meter operation?

A. At the lowest frequency of the channel

B. At the center frequency of the channel

C. At the highest frequency of the channel

D. On any frequency where the signal's sidebands are within the channel

B All amateur signals on 60 meter channels must be centered within the channel. [97.303(h)]

E1A07 Which amateur band requires transmission on specific channels rather than on a range of frequencies?

    A. 12 meter band
    B. 17 meter band
    C. 30 meter band
    D. 60 meter band

D Operation on 60 meters is restricted to five 2.8 kHz wide channels centered on 5332 kHz, 5348 kHz, 5368 kHz, 5373 kHz, and 5405 kHz with USB, data, and CW signals. This secondary allocation to amateurs is restricted in order to maintain compatibility with non-amateur stations who are the primary users of the band. [97.303(h)]

E1A08 If a station in a message forwarding system inadvertently forwards a message that is in violation of FCC rules, who is primarily accountable for the rules violation?

    A. The control operator of the packet bulletin board station
    B. The control operator of the originating station
    C. The control operators of all the stations in the system
    D. The control operators of all the stations in the system not authenticating the source from which they accept communications

B Amateur message systems are based on trusting that the original messages are proper and legal. Obviously control operators must be responsible for their communications. That's why the rules state, "For stations participating in a message forwarding system, the control operator of the station originating a message is primarily accountable for any violation of the rules in this Part contained in the message." The rules also state that the control operator of the first forwarding station must either authenticate the identity of the station from which it accepts a communication or accept accountability for any violation of the rules contained in messages it retransmits. [97.219(b), (d)]

E1A09 What is the first action you should take if your digital message forwarding station inadvertently forwards a communication that violates FCC rules?

A. Discontinue forwarding the communication as soon as you become aware of it
B. Notify the originating station that the communication does not comply with FCC rules
C. Notify the nearest FCC Field Engineer's office
D. Discontinue forwarding all messages

A   The FCC wants a problem fixed as quickly as possible. The rules say, "Except as noted in paragraph (d) of this section, for stations participating in a message forwarding system, the control operators of forwarding stations that retransmit inadvertently communications that violate the rules in this Part are not accountable for the violative communications. They are, however, responsible for discontinuing such communications once they become aware of their presence." [97.219(c)]

E1A10   If an amateur station is installed aboard a ship or aircraft, what condition must be met before the station is operated?

A. Its operation must be approved by the master of the ship or the pilot in command of the aircraft
B. The amateur station operator must agree not to transmit when the main radio of the ship or aircraft is in use
C. The amateur station must have a power supply that is completely independent of the main ship or aircraft power supply
D. The amateur operator must have an FCC Marine or Aircraft endorsement on his or her amateur license

A   The only additional requirement that the FCC imposes in this hypothetical situation is that the operation must be approved by the master of the ship. [97.11]

E1A11   Which of the following describes authorization or licensing required when operating an amateur station aboard a U.S.-registered vessel in international waters?

A. Any amateur license with an FCC Marine or Aircraft endorsement
B. Any FCC-issued amateur license
C. Only General class or higher amateur licenses
D. An unrestricted Radiotelephone Operator Permit

B   As a licensed US amateur you may operate with your full privileges from a US-registered vessel but only in international waters. Once in the territorial waters of another country, you are required to abide by their amateur licensing regulations. For the types of reciprocal permits, see Subelement E1F below. You are also required to have permission to operate. [97.5]

E1A12 With your transceiver displaying the carrier frequency of CW signals, you hear a DX station's CQ on 3.500 MHz. Is it legal to return the call using CW on the same frequency?

    A. Yes, the DX station initiated the contact

    B. Yes, the displayed frequency is within the 80 meter CW band segment

    C. No, one of the sidebands from the CW signal will be out of the band.

    D. No, U.S. stations are not permitted to use CW emissions below 3.525 MHz

C The sidebands of a CW signal extend above and below the frequency of the signal. See the discussion for E1A01 through E1A04.

E1A13 Who must be in physical control of the station apparatus of an amateur station aboard any vessel or craft that is documented or registered in the United States?

    A. Only a person with an FCC Marine Radio

    B. Any person holding an FCC issued amateur license or who is authorized for alien reciprocal operation

    C. Only a person named in an amateur station license grant

    D. Any person named in an amateur station license grant or a person holding an unrestricted Radiotelephone Operator Permit

B As a licensed US amateur you may operate with your full privileges from a US-registered vessel, but only in international waters. Once in the territorial waters of another country, you are required to abide by their amateur licensing regulations. For the types of reciprocal permits, see Subelement E1F below. You are also required to have permission to operate. [97.5]

E1A14 What is the maximum bandwidth for a data emission on 60 meters?

    A. 60 Hz

    B. 170 Hz

    C. 1.5 kHz

    D. 2.8 kHz

D Operation on 60 meters is restricted to five 2.8 kHz wide channels centered on 5332 kHz, 5348 kHz, 5368 kHz, 5373 kHz and 5405 kHz with USB, data, and CW signals. This secondary allocation to amateurs is restricted in order to maintain compatibility with non-amateur stations who are the primary users of the band. (See Figure E1A05.) [97.303(h)]

E1B   Station restrictions and special operations: restrictions on station location; general operating restrictions; spurious emissions; control operator reimbursement; antenna structure restrictions; RACES operations; national quiet zone

E1B01   Which of the following constitutes a spurious emission?

A. An amateur station transmission made at random without the proper call sign identification

B. A signal transmitted to prevent its detection by any station other than the intended recipient

C. Any transmitted signal that unintentionally interferes with another licensed radio station

D. An emission outside its necessary bandwidth that can be reduced or eliminated without affecting the information transmitted

D   Spurious emissions include harmonics, intermodulation or cross-modulation products, or distortion products. These signals are not required components of the desired signal and cause interference to other communications. [97.3(a)(43)]

E1B02   Which of the following factors might cause the physical location of an amateur station apparatus or antenna structure to be restricted?

A. The location is near an area of political conflict

B. The location is of geographical or horticultural importance

C. The location is in an ITU zone designated for coordination with one or more foreign governments

D. The location is of environmental importance or significant in American history, architecture, or culture

D   As you may already know, environmental or historical significance may limit land use in some cases. This includes where someone can install an antenna. That's why FCC reflects this concern in Part 97. [97.13(a)]

E1B03   Within what distance must an amateur station protect an FCC monitoring facility from harmful interference?

A. 1 mile

B. 3 miles

C. 10 miles

D. 30 miles

A   A 1600-meter (1-mile) protection zone is required to limit the effect of noise and other spurious emissions on sensitive facilities. [97.13(b)]

E1B04   What must be done before placing an amateur station within an officially designated wilderness area or wildlife preserve, or an area listed in the National Register of Historical Places?

A. A proposal must be submitted to the National Park Service
B. A letter of intent must be filed with the National Audubon Society
C. An Environmental Assessment must be submitted to the FCC
D. A form FSD-15 must be submitted to the Department of the Interior

C   You only have to deal with FCC. Fortunately, and to simplify matters, you do not have to deal with other governmental or non-governmental agencies. You'll also need to consult Part 1. [97.13(a), 1.1305 – 1.1319]

E1B05   What is the National Radio Quiet Zone?

A. An area in Puerto Rico surrounding the Arecibo Radio Telescope
B. An area in New Mexico surrounding the White Sands Test Area
C. An area surrounding the National Radio Astronomy Observatory
D. An area in Florida surrounding Cape Canaveral

C   The National Radio Quiet Zone protects the radio astronomy facilities for the National Radio Astronomy Observatory in Green Bank, West Virginia, and the Naval Research Laboratory at Sugar Grove, West Virginia. It is composed of parts of the states of Maryland, West Virginia and Virginia. [97.3(a)(33)]

E1B06   Which of the following additional rules apply if you are installing an amateur station antenna at a site at or near a public use airport?

A. You may have to notify the Federal Aviation Administration and register it with the FCC as required by Part 17 of FCC rules
B. No special rules apply if your antenna structure will be less than 300 feet in height
C. You must file an Environmental Impact Statement with the EPA before construction begins
D. You must obtain a construction permit from the airport zoning authority

A   Although most amateur exam questions seem to refer to Part 97, for this question you'll also need to consult Part 17. [97.15(a)]

E1B07   What is the highest modulation index permitted at the highest modulation frequency for angle modulation below 29 MHz?

A. 0.5
B. 1.0
C. 2.0
D. 3.0

B   Higher modulation indexes result in wider signal bandwidths than are necessary for good amateur practices. [97.307]

E1B08 What limitations may the FCC place on an amateur station if its signal causes interference to domestic broadcast reception, assuming that the receivers involved are of good engineering design?

A. The amateur station must cease operation
B. The amateur station must cease operation on all frequencies below 30 MHz
C. The amateur station must cease operation on all frequencies above 30 MHz
D. The amateur station must avoid transmitting during certain hours on frequencies that cause the interference

D Only by avoiding the frequency or frequencies used when the interference occurs would one reduce the interference and that's reflected in the rules. This assumes that the amateur station is not actually transmitting a spurious signal on the frequency of the broadcast station. [97.121(a)]

E1B09 Which amateur stations may be operated under RACES rules?

A. Only those club stations licensed to Amateur Extra class operators
B. Any FCC-licensed amateur station except a Technician class
C. Any FCC-licensed amateur station certified by the responsible civil defense organization for the area served
D. Any FCC-licensed amateur station participating in the Military Auxiliary Radio System (MARS)

C RACES operation requires proper registration and an amateur license of any grade. [97.3(a)(38), 97.407(a)]

E1B10 What frequencies are authorized to an amateur station participating under RACES rules?

A. All amateur service frequencies authorized to the control operator
B. Specific segments in the amateur service MF, HF, VHF and UHF bands
C. Specific local government channels
D. Military Auxiliary Radio System (MARS) channels

A The frequencies that can be used are determined by the control operator license. Under normal circumstances, there are no reserved frequencies for RACES operation. [97.407(b)]

E1B11 What is the permitted mean power of any spurious emission relative to the mean power of the fundamental emission from a station transmitter or external RF amplifier installed after January 1, 2003, and transmitting on a frequency below 30 MHz?

A. At least 43 dB below
B. At least 53 dB below
C. At least 63 dB below
D. At least 73 dB below

A   43 dB is equivalent to a ratio of 20,000 or approximately 7 S units. [97.307]

E1C   Definitions and restrictions pertaining to local, automatic and remote control operation; control operator responsibilities for remote and automatically controlled stations; IARP and CEPT licenses; third party communications over automatically controlled stations

E1C01 What is a remotely controlled station?

A. A station operated away from its regular home location
B. A station controlled by someone other than the licensee
C. A station operating under automatic control
D. A station controlled indirectly through a control link

D   All of these choices may sound like they could fit the definition, but they don't. The key phrase here is operation through a "control link" because the FCC definition is very specific. The rule says that remote control is "The use of a control operator who indirectly manipulates the operating adjustments in the station through a control link to achieve compliance with the FCC Rules." [97.3(a)(39)]

E1C02 What is meant by automatic control of a station?

A. The use of devices and procedures for control so that the control operator does not have to be present at a control point
B. A station operating with its output power controlled automatically
C. Remotely controlling a station's antenna pattern through a directional control link
D. The use of a control link between a control point and a locally controlled station

A   Like remote control, the FCC definition is very precise. The rule says, "The use of devices and procedures for control of a station when it is transmitting so that compliance with the FCC Rules is achieved without the control operator being present at a control point." Note that the compliance with FCC Rules is required. [97.3(a)(6), 97.109(d)]

E1C03 How do the control operator responsibilities of a station under automatic control differ from one under local control?

A. Under local control there is no control operator
B. Under automatic control the control operator is not required to be present at the control point
C. Under automatic control there is no control operator
D. Under local control a control operator is not required to be present at a control point

B By definition, the control operator of a station operating under automatic control is not required to be physically located at the control point. There must always be a control operator. Local control, as the name implies, requires that the control operator be present to manipulate the transmitter controls directly. [97.3(a)(6), 97.109]

E1C04 What is meant by IARP?

A. An international amateur radio permit that allows U.S. amateurs to operate in certain countries of the Americas
B. The internal amateur radio practices policy of the FCC
C. An indication of increased antenna reflected power.
D. A forecast of intermittent aurora radio propagation

A The IARP is the means by which the US and other countries in the Americas recognize each other's amateur licensees and allow them to operate when outside their home country.

E1C05 When may an automatically controlled station originate third party communications?

A. Never
B. Only when transmitting an RTTY or data emissions
C. When agreed upon by the sending and receiving stations
D. When approved by the National Telecommunication and Information Administration

A An automatically-controlled station may only respond to interrogation by a station under local or remote control. Since an automatically controlled digital station cannot initiate communications, "never" is the correct answer. [97.221(c)(1) and 97.115(c)]

E1C06   Which of the following statements concerning remotely controlled amateur stations is true?

   A. Only Extra Class operators may be the control operator of a remote station
   B. A control operator need not be present at the control point
   C. A control operator must be present at the control point
   D. Repeater and auxiliary stations may not be remotely controlled

C   Remote control is assumed to be identical to local control except that the operator is not at the station's control point which is where the functions of the station transmitter are adjusted. Unlike automatic control, the operator is still required to be in control of the transmitter when operating under remote control. [97.109(c)]

E1C07   What is meant by local control?

   A. Controlling a station through a local auxiliary link
   B. Automatically manipulating local station controls
   C. Direct manipulation of the transmitter by a control operator
   D. Controlling a repeater using a portable handheld transceiver

C   By definition, the control operator of a station operating under automatic control is not required to be physically located at the control point. There must always be a control operator. Local control, as the name implies, requires that the control operator be present to manipulate the transmitter controls directly. [97.3(a)(6), 97.3(a)(31),97.109]

E1C08   What is the maximum permissible duration of a remotely controlled station's transmissions if its control link malfunctions?

   A. 30 seconds
   B. 3 minutes
   C. 5 minutes
   D. 10 minutes

B   This is why many repeater time-out timers are set to 3 minutes! [97.213(b)]

E1C09   Which of these frequencies is available for an automatically controlled repeater operating below 30 MHz?

   A. 18.110 - 18.168 MHz
   B. 24.940 - 24.990 MHz
   C. 10.100 - 10.150 MHz
   D. 29.500 - 29.700 MHz

D   Ten meters is the only HF band on which any sort of repeater operation is permitted. [97.205(b)]

E1C10 What types of amateur stations may automatically retransmit the radio signals of other amateur stations?

A. Only beacon, repeater or space stations
B. Only auxiliary, repeater or space stations
C. Only earth stations, repeater stations or model craft
D. Only auxiliary, beacon or space stations

B Auxiliary stations are often used as remote repeater receivers, providing coverage in difficult terrain. Repeaters and space stations are designed to receive on one frequency, or one set of frequencies, then retransmit any signal they receive. Repeaters do this by demodulating the signal and routing the resulting audio or data through a separate transmitter. A space station may include a repeater or it may translate signals from one frequency to another without demodulating them. [97.113(d)]

E1C11 Which of the following operating arrangements allows an FCC-licensed U.S. citizen to operate in many European countries, and alien amateurs from many European countries to operate in the U.S.?

A. CEPT agreement
B. IARP agreement
C. ITU reciprocal license
D. All of these choices are correct

A CEPT is the European Conference of Posts and Telecommunications. The US is a participant in the CEPT Recommendation, which allows US amateurs to operate in certain European countries. It also allows amateur from many European countries to operate in the US. [97.5(d)]

E1C12 What types of communications may be transmitted to amateur stations in foreign countries?

A. Business-related messages for non-profit organizations
B. Messages intended for connection to users of the maritime satellite service
C. Communications incidental to the purpose of the amateur service and remarks of a personal nature
D. All of these choices are correct

C Even third-party communications are subject to the restrictions of non-commercial content. Although today the Internet provides an outlet for the disallowed types of communication, these rules were put in place at a time when few alternative services were available. It is still important to keep Amateur Radio free of commercial content. [97.117]

E1C13 Which of the following is required in order to operate in accordance with CEPT rules in foreign countries where permitted?

    A. You must identify in the official language of the country in which you are operating

    B. The U.S. embassy must approve of your operation

    C. You must bring a copy of FCC Public Notice DA 11-221

    D. You must append "/CEPT" to your call sign

C Similarly to carrying your amateur license, the CEPT treaty requires you to carry the FCC's Public Notice as documentation of the authorization to operate. It's also a good idea and may be required for you to carry your amateur license when operating.

E1D Amateur satellites: definitions and purpose; license requirements for space stations; available frequencies and bands; telecommand and telemetry operations; restrictions, and special provisions; notification requirements

E1D01 What is the definition of the term telemetry?

    A. One-way transmission of measurements at a distance from the measuring instrument

    B. Two-way radiotelephone transmissions in excess of 1000 feet

    C. Two-way single channel transmissions of data

    D. One-way transmission that initiates, modifies, or terminates the functions of a device at a distance

A This is a definition right out of the FCC rules, which define telemetry as, "A one-way transmission of measurements at a distance from the measuring instrument." [97.3(a)(46)] Practically, "measurements" include any kind of data from the system making the transmissions, such as repeater link status or error codes.

E1D02 What is the amateur satellite service?

    A. A radio navigation service using satellites for the purpose of self training, intercommunication and technical studies carried out by amateurs

    B. A spacecraft launching service for amateur-built satellites

    C. A radio communications service using amateur radio stations on satellites

    D. A radio communications service using stations on Earth satellites for public service broadcast

C The amateur satellite service is one of three Amateur Radio services mentioned in Part 97. [97.3(a)(2)] Radio amateurs have built and found launch opportunities for many satellites (called "space stations" in the rules). Transmissions to and from the satellites are carried out on specific sections of the amateur bands.

E1D03   What is a telecommand station in the amateur satellite service?

    A. An amateur station located on the Earth's surface for communication with other Earth stations by means of Earth satellites

    B. An amateur station that transmits communications to initiate, modify or terminate functions of a space station

    C. An amateur station located more than 50 km above the Earth's surface

    D. An amateur station that transmits telemetry consisting of measurements of upper atmosphere data

B   You need to know the definition for telecommand station, which is: "An amateur station that transmits communications to initiate, modify, or terminate functions of a space station." [97.3(a)(45)]

E1D04   What is an Earth station in the amateur satellite service?

    A. An amateur station within 50 km of the Earth's surface intended for communications with amateur stations by means of objects in space

    B. An amateur station that is not able to communicate using amateur satellites

    C. An amateur station that transmits telemetry consisting of measurement of upper atmosphere data

    D. Any amateur station on the surface of the Earth

A   The key is to notice that the questions deal with stations participating in the satellite service. An Earth station is an amateur station located on, or within 50 km of the Earth's surface intended for communications with space stations or with other Earth stations by means of one or more other objects in space. [97.3(a)(16)]

E1D05   What class of licensee is authorized to be the control operator of a space station?

    A. All except Technician Class

    B. Only General, Advanced or Amateur Extra Class

    C. Any class with appropriate operator privileges

    D. Only Amateur Extra Class

C   This may be surprising to you but any valid amateur license holder is eligible to be the control operator of a space station. However that individual is subject to the privileges of the class of operator license held. [97.207(a)]

E1D06   Which of the following is a requirement of a space station?

   A. The space station must be capable of terminating transmissions by
      telecommand when directed by the FCC
   B. The space station must cease all transmissions after 5 years
   C. The space station must be capable of changing its orbit whenever such a
      change is ordered by NASA
   D. All of these choices are correct

A   A space station must be capable of ceasing transmissions when ordered
to do so by the FCC. As a matter of interest, several amateur satellites have
enjoyed an operational life that exceeded 5 years. [97.207(b)]

E1D07   Which amateur service HF bands have frequencies authorized
for space stations?

   A. Only the 40 m, 20 m, 17 m, 15 m, 12 m and 10 m bands
   B. Only the 40 m, 20 m, 17 m, 15 m and 10 m bands
   C. Only the 40 m, 30 m, 20 m, 15 m, 12 m and 10 m bands
   D. All HF bands

A   Part 97 authorizes space station operation at HF in the 40, 20, 17, 15, 12,
and 10 meter bands. [97.207(c)]

E1D08   Which VHF amateur service bands have frequencies available for
space stations?

   A. 6 meters and 2 meters
   B. 6 meters, 2 meters, and 1.25 meters
   C. 2 meters and 1.25 meters
   D. 2 meters

D   Of the amateur VHF bands, only 2 meters has frequencies authorized for
use by space stations. [97.207(c)(2)]

E1D09   Which UHF amateur service bands have frequencies available for
a space station?

   A. Only 70 cm
   B. 70 cm and 13 cm
   C. 70 cm and 33 cm
   D. 33 cm and 13 cm

B   Now the question moves up to UHF where the 70, 23 and 13 cm bands
all have frequencies authorized for space stations. [97.207(c)(2)]

E1D10   Which amateur stations are eligible to be telecommand stations?

A. Any amateur station designated by NASA
B. Any amateur station so designated by the space station licensee, subject to the privileges of the class of operator license held by the control operator
C. Any amateur station so designated by the ITU
D. All of these choices are correct

B   The rule says, "Any amateur station designated by the licensee of a space station is eligible to transmit as a telecommand station for that space station, subject to the privileges of the class of operator license held by the control operator." [97.211(a)]

E1D11   Which amateur stations are eligible to operate as Earth stations?

A. Any amateur station whose licensee has filed a pre-space notification with the FCC's International Bureau
B. Only those of General, Advanced or Amateur Extra Class operators
C. Only those of Amateur Extra Class operators
D. Any amateur station, subject to the privileges of the class of operator license held by the control operator

D   Your license class is not an impediment to operate an Earth station. Any licensee can do it, limited to the privileges of the class of operator license held by the control operator. [97.209(a)]

E1E   Volunteer examiner program: definitions; qualifications; preparation and administration of exams; accreditation; question pools; documentation requirements

E1E01   What is the minimum number of qualified VEs required to administer an Element 4 amateur operator license examination?

A. 5
B. 2
C. 4
D. 3

D   The rules state, "Each examination for an amateur operator license must be administered by a team of at least 3 VEs at an examination session coordinated by a VEC." There may be more, of course. [97.509(a)]

E1E02   Where are the questions for all written U.S. amateur license examinations listed?

A. In FCC Part 97
B. In a question pool maintained by the FCC
C. In a question pool maintained by all the VECs
D. In the appropriate FCC Report and Order

C   The set of questions (one of which you're reading right now) is maintained by all the VECs in a common question pool. Each VEC may choose which questions it uses on exams from the pool (the ARRL VEC is currently using all of the questions), but all questions on the exams must be from the pool. [97.523]

E1E03   What is a Volunteer Examiner Coordinator?

A. A person who has volunteered to administer amateur operator license examinations
B. A person who has volunteered to prepare amateur operator license examinations
C. An organization that has entered into an agreement with the FCC to coordinate amateur operator license examinations
D. The person who has entered into an agreement with the FCC to be the VE session manager

C   A Volunteer Examiner Coordinator is an organization, not a person. The organization enters into an agreement with the FCC to be a VEC. The agreement with the FCC states that the organization exists to further the amateur service, is capable of acting as a VEC, will administer license exams for all classes, and will not discriminate against qualified candidates. [97.521]

E1E04   Which of the following best describes the Volunteer Examiner accreditation process?

A. Each General, Advanced and Amateur Extra Class operator is automatically accredited as a VE when the license is granted
B. The amateur operator applying must pass a VE examination administered by the FCC Enforcement Bureau
C. The prospective VE obtains accreditation from the FCC
D. The procedure by which a VEC confirms that the VE applicant meets FCC requirements to serve as an examiner

D   The procedure for accrediting a VE is created by each individual VEC. [97.509(b), 97.525]

E1E05 What is the minimum passing score on amateur operator license examinations?

A. Minimum passing score of 70%
B. Minimum passing score of 74%
C. Minimum passing score of 80%
D. Minimum passing score of 77%

B The minimum passing score for an Extra class exam is 37 questions out of 50. [97.503]

E1E06 Who is responsible for the proper conduct and necessary supervision during an amateur operator license examination session?

A. The VEC coordinating the session
B. The FCC
C. Each administering VE
D. The VE session manager

C The administering VEs are responsible for the proper conduct and necessary supervision of each examination. It is not just the responsibility of the VE in charge. [97.509(c)]

E1E07 What should a VE do if a candidate fails to comply with the examiner's instructions during an amateur operator license examination?

A. Warn the candidate that continued failure to comply will result in termination of the examination
B. Immediately terminate the candidate's examination
C. Allow the candidate to complete the examination, but invalidate the results
D. Immediately terminate everyone's examination and close the session

B The FCC is very explicit on this point. Discipline will be maintained. The rule says, "The administering VEs must immediately terminate the examination upon failure of the examinee to comply with their instructions." [97.509(c)]

E1E08 To which of the following examinees may a VE not administer an examination?

A. Employees of the VE
B. Friends of the VE
C. Relatives of the VE as listed in the FCC rules
D. All of these choices are correct

C Part 97 only restricts VEs from administering examinations to close relatives. It defines these as a spouse, children, grandchildren, stepchildren, parents, grandparents, stepparents, brothers, sisters, stepbrothers, stepsisters, aunts, uncles, nieces, nephews, and in-laws. [97.509(d)]

E1E09 What may be the penalty for a VE who fraudulently administers or certifies an examination?

    A. Revocation of the VE's amateur station license grant and the suspension of the VE's amateur operator license grant

    B. A fine of up to $1000 per occurrence

    C. A sentence of up to one year in prison

    D. All of these choices are correct

A   If you go to Part 97 you will find, "No VE may administer or certify any examination by fraudulent means or for monetary or other consideration including reimbursement in any amount in excess of that permitted. Violation of this provision may result in the revocation of the grant of the VE's amateur station license and the suspension of the grant of the VE's amateur operator license." [97.509(e)]

E1E10 What must the administering VEs do after the administration of a successful examination for an amateur operator license?

    A. They must collect and send the documents to the NCVEC for grading

    B. They must collect and submit the documents to the coordinating VEC for grading

    C. They must submit the application document to the coordinating VEC according to the coordinating VEC instructions

    D. They must collect and send the documents to the FCC according to instructions

C   The application document certifies a successful examination by the VE team. That document is then forwarded to the VEC according to the VEC procedures. [97.509(l)]

E1E11 What must the VE team do if an examinee scores a passing grade on all examination elements needed for an upgrade or new license?

    A. Photocopy all examination documents and forward them to the FCC for processing

    B. Three VEs must certify that the examinee is qualified for the license grant and that they have complied with the administering VE requirements

    C. Issue the examinee the new or upgrade license

    D. All these choices are correct

B   When the examinee has been credited for all examination elements required for the operator license sought, three VEs must certify that the examinee is qualified for the license grant and that the VEs have complied with the administering VE requirements. [97.509(h)]

E1E12 What must the VE team do with the application form if the examinee does not pass the exam?

    A. Return the application document to the examinee
    B. Maintain the application form with the VEC's records
    C. Send the application form to the FCC and inform the FCC of the grade
    D. Destroy the application form

A   The VE team must return the application form (such as the answer sheet) to the examinee and inform the examinee of the grade. The form may be kept by either the VE team or the examinee. VEC procedures may require the VE team to submit all non-passing application forms with the paperwork for the exam session. [97.509(j)]

E1E13 Which of these choices is an acceptable method for monitoring the applicants if a VEC opts to conduct an exam session remotely?

    A. Record the exam session on video tape for later examination by the VE team
    B. Use a real-time video link and the Internet to connect the exam session to the observing VEs
    C. The exam proctor observes the applicants and reports any violations
    D. Have each applicant sign an affidavit stating that all session rules were followed

B   Recognizing that travel time and expense can be an impediment for VEs and that the VE program has met high standards for integrity, the FCC recently decided to allow exam sessions to be monitored by VEs via a telecommunication link. The exam session must still be administered according to the VEC standards by on-site volunteers, whether they are VEs or not.

E1E14 For which types of out-of-pocket expenses do the Part 97 rules state that VEs and VECs may be reimbursed?

    A. Preparing, processing, administering and coordinating an examination for an amateur radio license
    B. Teaching an amateur operator license examination preparation course
    C. No expenses are authorized for reimbursement
    D. Providing amateur operator license examination preparation training materials

A   VEs cannot charge for exams. However, a fee may be charged to cover out-of-pocket and other administrative fees — a subtle difference. The rule says, "VEs and VECs may be reimbursed by examinees for out-of-pocket expenses incurred in preparing, processing, administering, or coordinating an examination for an amateur operator license." [97.527]

E1F Miscellaneous rules: external RF power amplifiers; business communications; compensated communications; spread spectrum; auxiliary stations; reciprocal operating privileges; special temporary authority

E1F01 On what frequencies are spread spectrum transmissions permitted?

A. Only on amateur frequencies above 50 MHz
B. Only on amateur frequencies above 222 MHz
C. Only on amateur frequencies above 420 MHz
D. Only on amateur frequencies above 144 MHz

B Because of the necessary bandwidth of spread spectrum signals, they are permitted only on the 222 MHz and higher frequency bands. [97.305]

E1F02 What privileges are authorized in the U.S. to persons holding an amateur service license granted by the government of Canada?

A. None, they must obtain a U.S. license
B. All privileges of the Extra Class license
C. The operating terms and conditions of the Canadian amateur service license, not to exceed U.S. Extra Class privileges
D. Full privileges, up to and including those of the Extra Class License, on the 80, 40, 20, 15, and 10 meter bands

C When Canadian amateurs operate from the US, they must operate according to the terms of their license as long as they do not exceed the privileges of the US Extra Class license. For example, Canadian hams visiting the US cannot operate on phone below 14.150 MHz.

E1F03 Under what circumstances may a dealer sell an external RF power amplifier capable of operation below 144 MHz if it has not been granted FCC certification?

A. It was purchased in used condition from an amateur operator and is sold to another amateur operator for use at that operator's station
B. The equipment dealer assembled it from a kit
C. It was imported from a manufacturer in a country that does not require certification of RF power amplifiers
D. It was imported from a manufacturer in another country, and it was certificated by that country's government

A There are three conditions that must be satisfied. First, the amplifier must be in used condition. Second, it must be purchased from an amateur operator and sold to another amateur operator. (The wording allows an equipment dealer to be involved.) Third, the amplifier must be for use at the purchasing operator's station. [97.315(b)(3)]

E1F04 Which of the following geographic descriptions approximately describes "Line A"?

A. A line roughly parallel to and south of the U.S.-Canadian border
B. A line roughly parallel to and west of the U.S. Atlantic coastline
C. A line roughly parallel to and north of the U.S.-Mexican border and Gulf coastline
D. A line roughly parallel to and east of the U.S. Pacific coastline

A If you operate on the 70 cm band in the northern states, you need to know about Line A which parallels the US-Canadian border. Line A exists to protect Canadian UHF non-amateurs using the 420 – 430 MHz band from interference by US amateur signals. The rules say, "No amateur station shall transmit from north of Line A in the 420-430 MHz segment." [97.3(a)(30), 97.303(m)(1)]

E1F05 Amateur stations may not transmit in which of the following frequency segments if they are located in the contiguous 48 states and north of Line A?

A. 440 - 450 MHz
B. 53 - 54 MHz
C. 222 - 223 MHz
D. 420 - 430 MHz

D See E1F04.

E1F06 Under what circumstances might the FCC issue a Special Temporary Authority (STA) to an amateur station?

A. To provide for experimental amateur communications
B. To allow regular operation on Land Mobile channels
C. To provide additional spectrum for personal use
D. To provide temporary operation while awaiting normal licensing

A While uncommon, STAs are granted in order for amateurs to experiment with unusual or new modes and on normally off-limit frequencies. For example, STAs were issued to allow amateurs to experiment with spread-spectrum techniques before a formal set of rules was put in place for all amateurs. [1.931]

E1F07   When may an amateur station send a message to a business?

    A. When the total money involved does not exceed $25

    B. When the control operator is employed by the FCC or another government agency

    C. When transmitting international third-party communications

    D. When neither the amateur nor his or her employer has a pecuniary interest in the communications

D   This falls into the area of prohibited transmissions. The rule prohibits, "Communications in which the station licensee or control operator has a pecuniary interest, including communications on behalf of an employer." [97.113(a)(3)]

E1F08   Which of the following types of amateur station communications are prohibited?

    A. Communications transmitted for hire or material compensation, except as otherwise provided in the rules

    B. Communications that have a political content, except as allowed by the Fairness Doctrine

    C. Communications that have a religious content

    D. Communications in a language other than English

A   With few exceptions, amateur communications must be free of any monetary interests. [97.113]

E1F09   Which of the following conditions apply when transmitting spread spectrum emission?

    A. A station transmitting SS emission must not cause harmful interference to other stations employing other authorized emissions

    B. The transmitting station must be in an area regulated by the FCC or in a country that permits SS emissions

    C. The transmission must not be used to obscure the meaning of any communication

    D. All of these choices are correct

D   All of the statements are correct. Here's the rule: "SS emission transmissions by an amateur station are authorized only for communications between points within areas where the amateur service is regulated by the FCC and between an area where the amateur service is regulated by the FCC and an amateur station in another country that permits such communications. SS emission transmissions must not be used for the purpose of obscuring the meaning of any communication." [97.311(a)]

E1F10 What is the maximum transmitter peak envelope power for an amateur station transmitting spread spectrum communications?

    A. 1 W
    B. 1.5 W
    C. 10 W
    D. 1.5 kW

C   The SS transmitter power must not exceed 10 W under any circumstances. [97.311(d)] This is to keep the noise from SS communications from causing interference to users of single-channel modes.

E1F11 Which of the following best describes one of the standards that must be met by an external RF power amplifier if it is to qualify for a grant of FCC certification?

    A. It must produce full legal output when driven by not more than 5 watts of mean RF input power
    B. It must be capable of external RF switching between its input and output networks
    C. It must exhibit a gain of 0 dB or less over its full output range
    D. It must satisfy the FCC's spurious emission standards when operated at the lesser of 1500 watts, or its full output power

D   The FCC rules state: "To receive a grant of certification, the amplifier must satisfy the spurious emission standards...when the amplifier is operated at the lesser of 1.5 kW PEP or its full output power and when the amplifier is placed in the 'standby' or 'off' positions while connected to the transmitter." [97.317(a)(1)]

E1F12 Who may be the control operator of an auxiliary station?

    A. Any licensed amateur operator
    B. Only Technician, General, Advanced or Amateur Extra Class operators
    C. Only General, Advanced or Amateur Extra Class operators
    D. Only Amateur Extra Class operators

B   Only Novice class licensees are not authorized to be the control operator of an auxiliary station. [97.201(a)]

Operating Procedures

There will be five questions on your Extra class examination from the Operating Procedures subelement. These five questions will be taken from the five groups of questions labeled E2A through E2E.

E2A Amateur radio in space: amateur satellites; orbital mechanics; frequencies and modes; satellite hardware; satellite operations; experimental telemetry applications

E2A01 What is the direction of an ascending pass for an amateur satellite?

A. From west to east
B. From east to west
C. From south to north
D. From north to south

C If the satellite is moving from south to north as it passes over your area, then it is making an ascending pass. Figure E2A01 illustrates basic satellite orbital terminology.

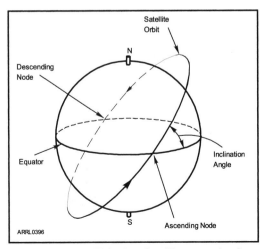

Figure E2A01 — This drawing illustrates basic satellite orbital terminology.

E2A02   What is the direction of a descending pass for an amateur satellite?

A. From north to south
B. From west to east
C. From east to west
D. From south to north

A   If the satellite is moving from north to south as it passes over your area, then it is making a descending pass. See Figure E2A01.

E2A03   What is the orbital period of an Earth satellite?

A. The point of maximum height of a satellite's orbit
B. The point of minimum height of a satellite's orbit
C. The time it takes for a satellite to complete one revolution around the Earth
D. The time it takes for a satellite to travel from perigee to apogee

C   The orbital period of a satellite is the time it takes that satellite to complete one orbit. See Figure E2A01.

E2A04   What is meant by the term mode as applied to an amateur radio satellite?

A. The type of signals that can be relayed through the satellite
B. The satellite's uplink and downlink frequency bands
C. The satellite's orientation with respect to the Earth
D. Whether the satellite is in a polar or equatorial orbit

B   The word "mode" is a bit overloaded when it comes to satellites. When referring to the satellite itself and not the type of signals it relays, mode refers to how the satellite's uplink and downlink frequencies are configured. Some satellites have only one mode, while others with multiple receivers and transmitters may be able to operate in several modes. Each frequency band is abbreviated by a single letter: VHF by V, UHF by U, L band by L, and so forth. The first letter specifies the uplink and the second letter specifies the downlink.

E2A05   What do the letters in a satellite's mode designator specify?

A. Power limits for uplink and downlink transmissions
B. The location of the ground control station
C. The polarization of uplink and downlink signals
D. The uplink and downlink frequency ranges

D   See E2A04.

Table E2-1
Satellite Operating Modes

Mode	Satellite Receive (Uplink)	Satellite Transmit (Downlink)
V/H	VHF (144 – 146 MHz)	HF (21 – 30 MHz)
U/V	UHF (435 – 438 MHz)	VHF (144 – 146 MHz)
V/U	VHF (144 – 146 MHz)	UHF (435 – 438 MHz)
L/U	L-Band (1.26 – 1.27 GHz)	UHF (435 – 438 MHz)

E2A06   On what band would a satellite receive signals if it were operating in mode U/V?

    A. 435 - 438 MHz
    B. 144 - 146 MHz
    C. 50.0 - 50.2 MHz
    D. 29.5 - 29.7 MHz

A   See E2A04.

E2A07   Which of the following types of signals can be relayed through a linear transponder?

    A. FM and CW
    B. SSB and SSTV
    C. PSK and Packet
    D. All of these choices are correct

D   By convention, transponder is the name given to a linear translator that is installed in a satellite. It is somewhat like a repeater in that both devices receive signals and retransmit them. A repeater does that for signals of a single mode on a single frequency. By contrast, a transponder's receive passband includes enough spectrum for many channels. The satellite transponder translates (or converts the frequency of) all signals in its passband — regardless of mode — and amplifies them and retransmits them in the new frequency range.

E2A08   Why should effective radiated power to a satellite which uses a linear transponder be limited?

A. To prevent creating errors in the satellite telemetry
B. To avoid reducing the downlink power to all other users
C. To prevent the satellite from emitting out-of-band signals
D. To avoid interfering with terrestrial QSOs

B   Satellites have a very limited power supply consisting of solar cells and batteries. This means that the satellite downlink (transmit) transmitter power must be limited. Most satellites use linear transponders (see E2A07) in which a strong signal on the uplink is also a strong signal on the downlink. If a satellite user's ERP is too high, it can consume a disproportionate amount of the available transmit power. For that reason, satellite users should use the minimum transmitter output power needed to communicate through the satellite.

E2A09   What do the terms L band and S band specify with regard to satellite communications?

A. The 23 centimeter and 13 centimeter bands
B. The 2 meter and 70 centimeter bands
C. FM and Digital Store-and-Forward systems
D. Which sideband to use

A   Instead of wavelength, microwave bands are designated by a letter. The Amateur Radio 23 cm band (1296 MHz) is part of the industry-designated L band. The Amateur Radio 13 cm band (2.4 GHz) is part of the industry-designated S band. (See Table E2-1.)

E2A10   Why may the received signal from an amateur satellite exhibit a rapidly repeating fading effect?

A. Because the satellite is spinning
B. Because of ionospheric absorption
C. Because of the satellite's low orbital altitude
D. Because of the Doppler Effect

A   Satellite designers usually spin a satellite to improve the stability of its orientation. Of course the satellite antennas spin too, and this results in a fairly rapid pulsed fading effect. This effect is called spin modulation.

E2A11 **What type of antenna can be used to minimize the effects of spin modulation and Faraday rotation?**

A. A linearly polarized antenna
B. A circularly polarized antenna
C. An isotropic antenna
D. A log-periodic dipole array

B See E2A10. Circularly polarized antennas of the proper sense (direction of polarization rotation) will minimize the effects of spin modulation. The polarization of a radio signal passing through the ionosphere does not remain constant. A horizontally polarized signal leaving a satellite will not be horizontally polarized after it passes through the ionosphere on its way to Earth. That signal will in fact seem to be changing polarization at a receiving station. This effect is called Faraday rotation. The best way to compensate for Faraday rotation is to use circularly polarized antennas for transmitting and receiving.

E2A12 **What is one way to predict the location of a satellite at a given time?**

A. By means of the Doppler data for the specified satellite
B. By subtracting the mean anomaly from the orbital inclination
C. By adding the mean anomaly to the orbital inclination
D. By calculations using the Keplerian elements for the specified satellite

D Johannes Kepler described the planetary orbits of our solar system. The laws and mathematical formulas that he developed may be used to calculate the location of a satellite at a given time. By using as input the values of a set of measurements describing the satellite orbit, called Keplerian elements, computer software can determine the location of the satellite.

E2A13 **What type of satellite appears to stay in one position in the sky?**

A. HEO
B. Geostationary
C. Geomagnetic
D. LEO

B The orbital period of a geosynchronous (geostationary) satellite is the same as the Earth's period of rotation about its axis. As a result, the satellite stays above the same spot on the Earth's surface, revolving around the Earth's axis at the same rate as the Earth's rotation.

E2A14   What technology is used to track, in real time, balloons carrying amateur radio transmitters?

    A. Radar
    B. Bandwidth compressed LORAN
    C. APRS
    D. Doppler shift of beacon signals

C   APRS (Automatic Packet Reporting System) packets containing latitude, longitude, and altitude information are generated by trackers (small automated stations) carried aloft by balloons. The same technology can be used to track radio-controlled craft.

E2B   Television practices: fast scan television standards and techniques; slow scan television standards and techniques

E2B01   How many times per second is a new frame transmitted in a fast-scan (NTSC) television system?

    A. 30
    B. 60
    C. 90
    D. 120

A   See E2B03.

E2B02   How many horizontal lines make up a fast-scan (NTSC) television frame?

    A. 30
    B. 60
    C. 525
    D. 1080

C   See E2B03.

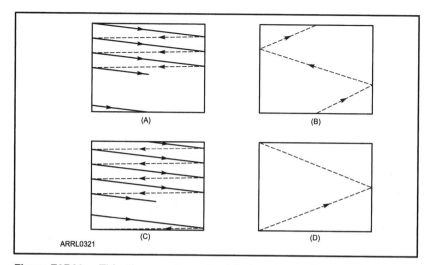

Figure E2B03 — This diagram shows the interlaced scanning used in analog TV. In field one, 262½ lines are scanned (A). At the end of field one, the electron scanning beam is returned to the top of the picture area (B). Scanning lines in field two (C) fall between the lines of field one. At the end of field two, the scanning beam is again returned to the top, where scanning continues with field one (D).

E2B03 How is an interlaced scanning pattern generated in a fast-scan (NTSC) television system?

A. By scanning two fields simultaneously
B. By scanning each field from bottom to top
C. By scanning lines from left to right in one field and right to left in the next
D. By scanning odd numbered lines in one field and even numbered ones in the next

D A picture is divided sequentially into pieces for transmission or viewing; this process is called scanning. A total of 525 scan lines comprise a frame (complete picture) in the US television system. Thirty frames are generated each second. The fast-scan TV frame consists of two fields of 262-1/2 lines each. The odd numbered lines are scanned in one field and the even numbered ones are scanned in the next. The half line is the secret to the interlaced scan pattern.

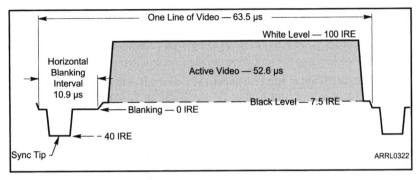

Figure E2B04 — The RS-170 monochrome waveform for one video line. A full NTSC video frame consists of 525 video lines organized as two interlaced fields. The horizontal blanking interval shown here turns off the scanning beam as the beam returns to the start of the next line. The vertical blanking interval (not shown) turns off the beam while it returns to the top of the screen.

E2B04   What is blanking in a video signal?

A. Synchronization of the horizontal and vertical sync pulses
B. Turning off the scanning beam while it is traveling from right to left or from bottom to top
C. Turning off the scanning beam at the conclusion of a transmission
D. Transmitting a black and white test pattern

B   The function of blanking is to turn off the scanning beam while the beam is traveling from right to left and from bottom to top. The blanking signal occurs at the end of each scan line and at the end of each scan field. In terms of video brightness, blanking is blacker than black so that the blanked beam will not affect the image as it travels across it.

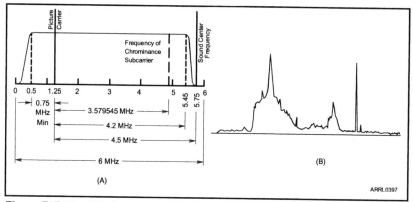

Figure E2B05 — The frequency spectrum of a color fast-scan TV signal is shown in part A. The vestigial sideband is to the left of the line labeled "Picture Carrier" and occupies a bandwidth of 1.25 MHz. Part B shows the spectrum analyzer view of a typical fast-scan TV signal.

E2B05 Which of the following is an advantage of using vestigial sideband for standard fast-scan TV transmissions?

A. The vestigial sideband carries the audio information
B. The vestigial sideband contains chroma information
C. Vestigial sideband reduces bandwidth while allowing for simple video detector circuitry
D. Vestigial sideband provides high frequency emphasis to sharpen the picture

C The frequency spectrum of a color fast-scan TV signal is shown in part A of Figure E2B05. The vestigial sideband is to the left of the line labeled Picture Carrier and occupies a bandwidth of 1.25 MHz. Part B shows the spectrum analyzer view of a typical fast-scan TV signal. Including a vestigial sideband (see the next question) allows the video signal to be recovered with simple AM detector circuits, as opposed to requiring the more complex SSB demodulators. By only including part of the unused sideband, the full bandwidth of an AM-DSB signal is not required.

E2B06 What is vestigial sideband modulation?

    A. Amplitude modulation in which one complete sideband and a portion of the other are transmitted

    B. A type of modulation in which one sideband is inverted

    C. Narrow-band FM transmission achieved by filtering one sideband from the audio before frequency modulating the carrier

    D. Spread spectrum modulation achieved by applying FM modulation following single sideband amplitude modulation

A See E2B05. Vestigial means partial and unused, and the vestigial sideband of a fast-scan TV signal is attenuated and reduced in bandwidth from the full-bandwidth sideband that carries the video information. Its only function is to allow the transmitted signal to be detected as an AM signal, instead of SSB.

E2B07 What is the name of the signal component that carries color information in NTSC video?

    A. Luminance

    B. Chroma

    C. Hue

    D. Spectral Intensity

B Also referred to as chrominance, the chroma signal is combined with the basic monochrome (black and white) TV signal.

E2B08 Which of the following is a common method of transmitting accompanying audio with amateur fast-scan television?

    A. Frequency-modulated sub-carrier

    B. A separate VHF or UHF audio link

    C. Frequency modulation of the video carrier

    D. All of these choices are correct

D The first choice is how commercial stations transmit the audio information in a TV signal. The second is the way most hams send ATV audio because it is easier than adding an FM signal to the transmitted video signal. The third choice is another popular method of adding audio information without affecting the AM video components

E2B09 What hardware, other than a receiver with SSB capability and a suitable computer, is needed to decode SSTV using Digital Radio Mondiale (DRM)?

A. A special IF converter
B. A special front end limiter
C. A special notch filter to remove synchronization pulses
D. No other hardware is needed

D Digital Radio Mondiale was developed for transmitting digital audio information and data over shortwave broadcast channels. Hams adapted it to transmit SSTV pictures as digital data. The demodulated audio from a DRM signal can be processed by computer sound card, recovering the picture information.

E2B10 Which of the following is an acceptable bandwidth for Digital Radio Mondiale (DRM) based voice or SSTV digital transmissions made on the HF amateur bands?

A. 3 kHz
B. 10 kHz
C. 15 kHz
D. 20 kHz

A Image transmissions are allowed in phone band segments if their bandwidth is no greater than that of a voice signal of the same modulation type. For SSB transmissions, the normal bandwidth is 3 kHz.

E2B11 What is the function of the Vertical Interval Signaling (VIS) code sent as part of an SSTV transmission?

A. To lock the color burst oscillator in color SSTV images
B. To identify the SSTV mode being used
C. To provide vertical synchronization
D. To identify the call sign of the station transmitting

B There are a number of different formats in which SSTV images are transmitted. So that automated receiving systems can determine which format is being used, identifying codes are transmitted during the vertical blanking period between images. This is called Vertical Interval Signaling (VIS).

E2B12 How are analog SSTV images typically transmitted on the HF bands?

    A. Video is converted to equivalent Baudot representation
    B. Video is converted to equivalent ASCII representation
    C. Varying tone frequencies representing the video are transmitted using PSK
    D. Varying tone frequencies representing the video are transmitted using single sideband

D The standard method of transmitting analog SSTV on the HF bands uses an SSB transmitter. The raster control signals, such as horizontal and vertical sync pulses, and image video are encoded as different tones that are transmitted as regular audio over the SSB channel.

E2B13 How many lines are commonly used in each frame on an amateur slow-scan color television picture?

    A. 30 to 60
    B. 60 or 100
    C. 128 or 256
    D. 180 or 360

C Color SSTV standards specify either 128 or 256 lines per frame depending on the selected transmission format.

E2B14 What aspect of an amateur slow-scan television signal encodes the brightness of the picture?

    A. Tone frequency
    B. Tone amplitude
    C. Sync amplitude
    D. Sync frequency

A Because HF signals experience frequent fading, amplitude is not used to encode the SSTV signal. Frequency of the received tones is unaffected by fading, however, and can be used to encode brightness or any control function of the signal.

E2B15 What signals SSTV receiving equipment to begin a new picture line?

    A. Specific tone frequencies
    B. Elapsed time
    C. Specific tone amplitudes
    D. A two-tone signal

A The standard method of transmitting analog SSTV on the HF bands uses an SSB transmitter. The raster control signals, such as horizontal and vertical sync pulses, and image video are encoded as different tones that are transmitted as regular audio over the SSB channel.

E2B16 Which is a video standard used by North American Fast Scan ATV stations?

    A. PAL
    B. DRM
    C. Scottie
    D. NTSC

D NTSC stands for National Television System Committee and is the standard for analog commercial television broadcasters, as well. The NTSC standard describes all aspects of the television signal.

E2B17 What is the approximate bandwidth of a slow-scan TV signal?

    A. 600 Hz
    B. 3 kHz
    C. 2 MHz
    D. 6 MHz

B Image transmissions are allowed in phone band segments if their bandwidth is no greater than that of a voice signal of the same modulation type. For SSB transmissions on HF, the normal bandwidth is 3 kHz.

E2B18 On which of the following frequencies is one likely to find FM ATV transmissions?

    A. 14.230 MHz
    B. 29.6 MHz
    C. 52.525 MHz
    D. 1255 MHz

D Because the bandwidth of FM ATV signals is so wide (approximately 12 MHz), its use is restricted to UHF and higher bands.

E2B19 What special operating frequency restrictions are imposed on slow scan TV transmissions?

    A. None; they are allowed on all amateur frequencies
    B. They are restricted to 7.245 MHz, 14.245 MHz, 21.345 MHz, and 28.945 MHz
    C. They are restricted to phone band segments and their bandwidth can be no greater than that of a voice signal of the same modulation type
    D. They are not permitted above 54 MHz

C Image transmissions are allowed in phone band segments if their bandwidth is no greater than that of a voice signal of the same modulation type. For SSB transmissions on HF, the normal bandwidth is 3 kHz.

E2C Operating methods: contest and DX operating; remote operation techniques; Cabrillo format; QSLing; RF network connected systems

E2C01   Which of the following is true about contest operating?

A. Operators are permitted to make contacts even if they do not submit a log
B. Interference to other amateurs is unavoidable and therefore acceptable
C. It is mandatory to transmit the call sign of the station being worked as part of every transmission to that station
D. Every contest requires a signal report in the exchange

A   With very few exceptions, contests are open to all amateurs.. In addition, stations participating in the contest may count contacts with you even if you do not send the event sponsor a log of your contest activity. Just ask the contest station what information they need to complete the contact and they'll tell you.

E2C02   Which of the following best describes the term self-spotting in regards to contest operating?

A. The generally prohibited practice of posting one's own call sign and frequency on a spotting network
B. The acceptable practice of manually posting the call signs of stations on a spotting network
C. A manual technique for rapidly zero beating or tuning to a station's frequency before calling that station
D. An automatic method for rapidly zero beating or tuning to a station's frequency before calling that station

A   A "spot" is an announcement that a specific station is present on a specific frequency. Spots are distributed mostly over the Internet via websites and TELNET-based systems that form a worldwide spotting network. Posting your own frequency and call sign — self-spotting — in an attempt to get more contacts is prohibited.

E2C03   From which of the following bands is amateur radio contesting generally excluded?

A. 30 m
B. 6 m
C. 2 m
D. 33 cm

A   By general agreement, 60, 30, 17 and 12 meters are excluded from contest activity. This gives stations not interested in the contest an additional option for avoiding contest activity.

E2C04 What type of transmission is most often used for a ham radio mesh network?

 A. Spread spectrum in the 2.4 GHz band
 B. Multiple Frequency Shift Keying in the 10 GHz band
 C. Store and forward on the 440 MHz band
 D. Frequency division multiplex in the 24 GHz band

A   Ham mesh networks are often based on wireless networking WiFi routers that have been re-programmed to operate on the 2.4 GHz band channels shared with amateurs. Once restricted to amateur frequencies, the router can implement a wide variety of networking protocols.

E2C05 What is the function of a DX QSL Manager?

 A. To allocate frequencies for DXpeditions
 B. To handle the receiving and sending of confirmation cards for a DX station
 C. To run a net to allow many stations to contact a rare DX station
 D. To relay calls to and from a DX station

B   Because of the expense of postage and the effort required to answer a lot of QSL requests, many DX stations enlist the aid of a manager in the US, Europe, Japan or another area with good postal service. QSL cards are sent directly to the manager without the DX station having to take time out from operating. This practice generally saves on postage for the station sending the confirmation, as well.

E2C06 During a VHF/UHF contest, in which band segment would you expect to find the highest level of activity?

 A. At the top of each band, usually in a segment reserved for contests
 B. In the middle of each band, usually on the national calling frequency
 C. In the weak signal segment of the band, with most of the activity near the calling frequency
 D. In the middle of the band, usually 25 kHz above the national calling frequency

C   Most contest activity on the VHF and UHF bands is SSB or CW, referred to as weak signal because these modes perform better than FM at low signal-to-noise ratios. The usual convention for contest activity is to call CQ on or near the calling frequency and then move to a nearby frequency once activity builds. In populated areas, calling CQ on the calling frequency is discouraged so that stations will spread out.

E2C07   What is the Cabrillo format?

   A. A standard for submission of electronic contest logs
   B. A method of exchanging information during a contest QSO
   C. The most common set of contest rules
   D. The rules of order for meetings between contest sponsors

A   Named for Cabrillo College near Santa Cruz, California, the Cabrillo format standard specifies how log information in text form is organized for submission to the contest sponsor. It does not control how logs are checked or scored by the sponsor.

E2C08   Which of the following contacts may be confirmed through the U.S. QSL bureau system?

   A. Special event contacts between stations in the U.S.
   B. Contacts between a U.S. station and a non-U.S. station
   C. Repeater contacts between U.S. club members
   D. Contacts using tactical call signs

B   The bureau system would be overwhelmed if US-to-US QSLs so it is limited to the exchange of U.S. and DX QSLs. The ARRL's electronic Logbook of the World (LoTW) system accepts QSOs between any amateur stations.

E2C09   What type of equipment is commonly used to implement a ham radio mesh network?

   A. A 2 meter VHF transceiver with a 1200 baud modem
   B. An optical cable connection between the USB ports of 2 separate computers
   C. A standard wireless router running custom software
   D. A 440 MHz transceiver with a 9600 baud modem

C   See E2C04

E2C10   Why might a DX station state that they are listening on another frequency?

   A. Because the DX station may be transmitting on a frequency that is prohibited to some responding stations
   B. To separate the calling stations from the DX station
   C. To improve operating efficiency by reducing interference
   D. All of these choices are correct

D   By transmitting on one frequency and listening on another (split frequency operation) when many stations are calling, it is much easier for the callers to hear the DX station. This allows everyone to follow instructions and makes the operation much more efficient. If you hear a DX station working callers you can't hear, tune around to see if you can find the callers on a nearby frequency — usually above that of the DX station.

E2C11 How should you generally identify your station when attempting to contact a DX station during a contest or in a pileup?

A. Send your full call sign once or twice
B. Send only the last two letters of your call sign until you make contact
C. Send your full call sign and grid square
D. Send the call sign of the DX station three times, the words "this is", then your call sign three times

A When the other operator sends CQ or QRZ? (Who is calling me?), this is your cue to call by sending your full call sign. Sending only part of your call sign is not a proper way to identify your station.

E2C12 What might help to restore contact when DX signals become too weak to copy across an entire HF band a few hours after sunset?

A. Switch to a higher frequency HF band
B. Switch to a lower frequency HF band
C. Wait 90 minutes or so for the signal degradation to pass
D. Wait 24 hours before attempting another communication on the band

B Weak and fluttery signals after sunset indicate that the band conditions are deteriorating. You can continue to operate by changing frequencies. In this case, you'll want to operate at a lower frequency because the MUF has moved lower as well.

E2C13 What indicator is required to be used by U.S.-licensed operators when operating a station via remote control where the transmitter is located in the U.S.?

A. / followed by the USPS two letter abbreviation for the state in which the remote station is located
B. /R# where # is the district of the remote station
C. The ARRL section of the remote station
D. No additional indicator is required

D The same rules for identification apply to stations operating under local control and remote control.

E2D Operating methods: VHF and UHF digital modes and procedures; APRS; EME procedures, meteor scatter procedure

E2D01   Which of the following digital modes is especially designed for use for meteor scatter signals?

A. WSPR
B. FSK441
C. Hellschreiber
D. APRS

B   The WSJT suite of software written by Joe Taylor, K1JT includes several digital modes optimized for special operating modes such as moonbounce (EME), meteor scatter, low-power beacons, and more. FSK441 uses multi-frequency-shift keying using four tones at a data rate of 441 baud.

E2D02   Which of the following is a good technique for making meteor-scatter contacts?

A. 15 second timed transmission sequences with stations alternating based on location
B. Use of high speed CW or digital modes
C. Short transmission with rapidly repeated call signs and signal reports
D. All of these choices are correct

D   The short life of meteor trails means that stations must exchanged data very rapidly and can't spend a lot of time calling each other. As in moonbounce communication, one solution is to synchronize calling and listening periods. Another solution is to use modes that support high-speed data transmission and keep transmit times very short. (See also E2D01)

E2D03   Which of the following digital modes is especially useful for EME communications?

A. FSK441
B. PACTOR III
C. Olivia
D. JT65

D   The WSJT suite of software written by Joe Taylor, K1JT includes several digital modes optimized for special operating modes such as moonbounce (EME), meteor scatter, low-power beacons, and more. JT65 is optimized for use at the extremely low signal-to-noise ratios and long transit times encountered in EME communications.

E2D04   What is the purpose of digital store-and-forward functions on an Amateur Radio satellite?

A. To upload operational software for the transponder
B. To delay download of telemetry between satellites
C. To store digital messages in the satellite for later download by other stations
D. To relay messages between satellites

C   Digital communications satellites provide non-real-time computer-to-computer communications. Since satellites in low Earth orbit (LEO) cannot see large segments of the world at once, these satellites work like temporary mailboxes in space. You upload a message or a file to a satellite and it is stored for a time (could be days or weeks) until someone else — possibly on the other side of the world — downloads it. Messages or files can be sent to an individual recipient or everyone. Such communications are called store-and-forward.

E2D05   Which of the following techniques is normally used by low Earth orbiting digital satellites to relay messages around the world?

A. Digipeating
B. Store-and-forward
C. Multi-satellite relaying
D. Node hopping

B   See E2D04.

E2D06   Which of the following describes a method of establishing EME contacts?

A. Time synchronous transmissions alternately from each station
B. Storing and forwarding digital messages
C. Judging optimum transmission times by monitoring beacons from the Moon
D. High speed CW identification to avoid fading

A   For analog or digital mode EME contacts, the round-trip time and extremely weak signals make the usual call-and-answer method impractical. The current standard method is for stations to call at synchronized times so that it is clear when to listen and when to transmit.

E2D07   What digital protocol is used by APRS?

A. PACTOR
B. 802.11
C. AX.25
D. AMTOR

C   The APRS system is constructed to use the amateur AX.25 packet protocol.

E2D08 What type of packet frame is used to transmit APRS beacon data?

A. Unnumbered Information
B. Disconnect
C. Acknowledgement
D. Connect

A The APRS system is constructed to use the amateur AX.25 packet protocol. To simplify the communication between stations, an APRS beacon is transmitted as an unnumbered information (UI) frame. The stations in an APRS network are not connected in the normal packet radio sense. The beacon frames are not directed to a specific station, and receiving stations do not acknowledge correct receipt of the frames.

E2D09 Which of these digital modes has the fastest data throughput under clear communication conditions?

A. AMTOR
B. 170 Hz shift, 45 baud RTTY
C. PSK31
D. 300 baud packet

D For fastest throughput under clear communications conditions (assumed to be error-free transmission) the highest data rate is needed. Of the choices given, 300-baud packet has the highest data rate and, therefore, the fastest data throughput of these choices.

E2D10 How can an APRS station be used to help support a public service communications activity?

A. An APRS station with an emergency medical technician can automatically transmit medical data to the nearest hospital
B. APRS stations with General Personnel Scanners can automatically relay the participant numbers and time as they pass the check points
C. An APRS station with a GPS unit can automatically transmit information to show a mobile station's position during the event
D. All of these choices are correct

C In the Automatic Packet Reporting System, positions of fixed stations can be entered directly into software. Mobile APRS stations equipped with a GPS (Global Positioning System) unit can automatically transmit information to show the station's position. This capability can be useful in support of a public service communications activity, such as a walk-a-thon.

E2D11 Which of the following data are used by the APRS network to communicate your location?

A. Polar coordinates
B. Time and frequency
C. Radio direction finding spectrum analysis
D. Latitude and longitude

D Many TNCs used for APRS have a serial port that can be connected directly to the data output jack of a GPS receiver. Software in the TNC recognizes the NMEA-0183 format and automatically extracts the latitude and longitude information for transmission over the APRS network. If the GPS receiver is not connected directly to the TNC, an operator can enter the information manually.

E2D12 How does JT65 improve EME communications?

A. It can decode signals many dB below the noise floor using FEC
B. It controls the receiver to track Doppler shift
C. It supplies signals to guide the antenna to track the Moon
D. All of these choices are correct

A The forward error correcting codes (FEC) used by JT65 and precisely controlled timing and frequencies of the signals allows JT65 processing software to recover very weak signals error-free. Using JT65 allows much smaller stations to make EME contacts than if they were to use an analog mode such as CW or SSB that are copied by ear.

E2D13 What type of modulation is used for JT65 contacts?

A. Multi-tone AFSK
B. PSK
C. RTTY
D. IEEE 802.11

A The sequence of tones is carefully encoded, enabling a receiver to extract the signal from deep below the noise level.

E2D14 What is one advantage of using JT65 coding?

A. Uses only a 65 Hz bandwidth
B. The ability to decode signals which have a very low signal to noise ratio
C. Easily copied by ear if necessary
D. Permits fast-scan TV transmissions over narrow bandwidth

B JT65 is an example of a protocol and modulation technique that uses repeated and predictable sequences of data to be combined in a way that diminishes the effect of noise. This is called processing gain and it allows signals to be demodulated with signal levels well below that of the noise.

E2E   Operating methods: operating HF digital modes

E2E01   Which type of modulation is common for data emissions below 30 MHz?

    A. DTMF tones modulating an FM signal
    B. FSK
    C. Pulse modulation
    D. Spread spectrum

B   The most common method of transmitting data emissions below 30 MHz is by FSK (frequency-shift keying) of an RF carrier using an SSB transceiver.

E2E02   What do the letters FEC mean as they relate to digital operation?

    A. Forward Error Correction
    B. First Error Correction
    C. Fatal Error Correction
    D. Final Error Correction

A   The telecommunications industry uses FEC as a standard acronym for Forward Error Correction. FEC involves the use of sending redundant data or special codes with the data so that the receiving system can compensate for errors that occur during transmission and reception

E2E03   How is the timing of JT65 contacts organized?

    A. By exchanging ACK/NAK packets
    B. Stations take turns on alternate days
    C. Alternating transmissions at 1 minute intervals
    D. It depends on the lunar phase

C   The protocol uses precise timing between the transmitting and receiving stations and extensive codes to allow the received signal to gradually be separated from noise. JT65 uses 1-minute alternating transmit-receive periods synchronized to accurate timing signals. See E2D14.

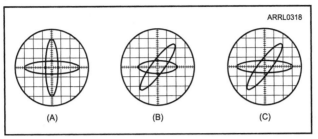

Figure E2E04 — The two tones of an FSK or AFSK signal are represented as a pair of ellipses on a crossed-ellipse display. For the best decoding, the signal should be tuned so that the ellipses are of equal size and at right angles as in (A). Displays such as at (B) and (C) indicate a mistuned signal.

E2E04   What is indicated when one of the ellipses in an FSK crossed-ellipse display suddenly disappears?

A. Selective fading has occurred
B. One of the signal filters has saturated
C. The receiver has drifted 5 kHz from the desired receive frequency
D. The mark and space signal have been inverted

A   A common tuning aid for digital FSK signals uses a crossed pair of ellipses. Each ellipse represents one of the two tones being sent. With the propagation effect known as selective fading, a small region of the radio spectrum undergoes a deep fade as if a narrow filter has been applied to it. This is what can make one of the ellipses on the terminal unit display disappear.

E2E05   Which type of digital mode does not support keyboard-to-keyboard operation?

A. Winlink
B. RTTY
C. PSK31
D. MFSK

A   Keyboard-to-keyboard operation refers to each character being transmitted and displayed as it is typed (RTTY, PSK31, MFSK) or after an end-of-line character is entered by the operator (Packet). Winlink is an email system that transmits entire files that are not displayed until the receiving operator opens and reads them.

E2E06   What is the most common data rate used for HF packet?

A. 48 baud
B. 110 baud
C. 300 baud
D. 1200 baud

C   The most common data rate used for HF packet communications is 300 baud. Slower signaling rates are used for RTTY and higher rates are not permitted by regulation because they require more bandwidth.

E2E07   What is the typical bandwidth of a properly modulated MFSK16 signal?

A. 31 Hz
B. 316 Hz
C. 550 Hz
D. 2.16 kHz

B   MFSK16 uses 16 tones, which are sent one at a time at 15.625 baud. Because the 16 tones are spaced 15.625 Hz apart, a properly modulated MFSK16 signal has a bandwidth of 316 Hz.

E2E08   Which of the following HF digital modes can be used to transfer binary files?

A. Hellschreiber
B. PACTOR
C. RTTY
D. AMTOR

B   PACTOR was developed to combine the best features of packet radio and AMTOR. Since PACTOR carries binary data, you can use this mode to transfer binary data files directly between stations. The other mode choices are not suitable for this purpose.

E2E09   Which of the following HF digital modes uses variable-length coding for bandwidth efficiency?

A. RTTY
B. PACTOR
C. MT63
D. PSK31

D   Peter Martinez, G3PLX, invented Varicode as a way to improve the data rate of PSK31. In Varicode, the more common characters have shorter codes, just like in Morse code.

E2E10 Which of these digital communications modes has the narrowest bandwidth?

A. MFSK16
B. 170-Hz shift, 45 baud RTTY
C. PSK31
D. 300-baud packet

C PSK31 uses a data rate of 31.25 baud. This results in an RF signal that has a bandwidth of only 37.5 Hz! This is by far the narrowest bandwidth of any amateur digital communications mode.

E2E11 What is the difference between direct FSK and audio FSK?

A. Direct FSK applies the data signal to the transmitter VFO
B. Audio FSK has a superior frequency response
C. Direct FSK uses a DC-coupled data connection
D. Audio FSK can be performed anywhere in the transmit chain

A With the advent of direct digital synthesis (DDS) VFOs, many radios now feature a direct FSK input by which a digital signal can shift the VFO frequency directly. This often results in superior modulation characteristics that reduce data errors by the receiver.

E2E12 Which type of control is used by stations using the Automatic Link Enable (ALE) protocol?

A. Local
B. Remote
C. Automatic
D. ALE can use any type of control

C Because the Automatic Link Establishment (ALE) protocol switches frequencies and initiates transmissions automatically, this is an example of automatic control.

E2E13 Which of the following is a possible reason that attempts to initiate contact with a digital station on a clear frequency are unsuccessful?

A. Your transmit frequency is incorrect
B. The protocol version you are using is not supported by the digital station
C. Another station you are unable to hear is using the frequency
D. All of these choices are correct

D Each aspect of the transmission must be compatible with the receiver, beginning with the transmit frequency. The hidden transmitter problem can also cause a communication link to fail because of interference.

Radio Wave Propagation

There will be three questions on your Extra class examination from the Radio Wave Propagation subelement. These three questions will be taken from the three groups of questions labeled E3A through E3C.

E3A   Electromagnetic waves; Earth-Moon-Earth communications; meteor scatter; microwave tropospheric and scatter propagation; aurora propagation

E3A01   What is the approximate maximum separation measured along the surface of the Earth between two stations communicating by moonbounce?

    A. 500 miles, if the Moon is at perigee
    B. 2000 miles, if the Moon is at apogee
    C. 5000 miles, if the Moon is at perigee
    D. 12,000 miles, if the Moon is visible by both stations

D   The Moon must be visible to both stations at the same time in order for them to communicate by reflecting VHF or UHF signals off the lunar surface. Those stations may be separated by nearly 180° of arc on the Earth's surface — a distance of more than 11,000 miles. There is no specific maximum distance between two stations to communicate via moonbounce, as long as they have a mutual lunar window. In other words, the moon must be above the radio horizon where both stations can see it at the same time.

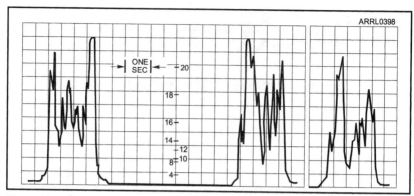

Figure E3A02 — This graph is a chart recording showing the deep libration fading of echoes received at W2NFA from the Moon's surface.

E3A02 What characterizes libration fading of an EME signal?

A. A slow change in the pitch of the CW signal
B. A fluttery irregular fading
C. A gradual loss of signal as the Sun rises
D. The returning echo is several hertz lower in frequency than the transmitted signal

B Libration fading is multipath scattering of the radio waves from the very large (2000-mile diameter) and rough Moon surface combined with the relatively short-term variations of the Moon in its orbit. Libration fading of an EME signal is characterized in general as fluttery, rapid, irregular fading not unlike that observed in tropospheric-scatter propagation. Fading can be very deep, 20 dB or more, and the maximum fading will depend on the operating frequency. You can see the effects of libration fading in the accompanying figure recorded at the station of W2NFA.

E3A03 When scheduling EME contacts, which of these conditions will generally result in the least path loss?

A. When the Moon is at perigee
B. When the Moon is full
C. When the Moon is at apogee
D. When the MUF is above 30 MHz

A The Moon's orbit is slightly elliptical, with the closest distance (perigee) being 225,000 miles and the furthest (apogee) being 252,000 miles. EME path loss is typically 2 dB less at perigee.

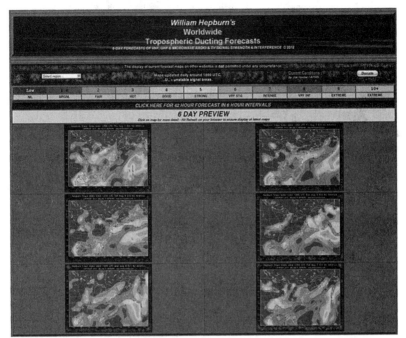

Figure E3A04 — Hepburn maps are used by VHF+ operators to look for possible tropospheric propagation.

E3A04   What do Hepburn maps predict?

A. Sporadic E propagation
B. Locations of auroral reflecting zones
C. Likelihood of rain scatter along cold or warm fronts
D. Probability of tropospheric propagation

D   Tropospheric propagation often occurs along variations in humidity or temperature in the atmosphere. Hepburn maps predict the likelihood of these variations being strong enough to support radio wave propagation.

E3A05   Tropospheric propagation of microwave signals often occurs along what weather related structure?

A. Gray-line
B. Lightning discharges
C. Warm and cold fronts
D. Sprites and jets

C   At times, weather conditions such as temperature inversions and warm or cold fronts can create sharp transitions between air layers. These transitions can reflect or guide VHF, UHF, and microwave radio waves, forming *ducts* in the troposphere (lower layers of the atmosphere), similar to propagation in a waveguide. This form of propagation is called *tropospheric ducting*.

E3A06   Which of the following is required for microwave propagation via rain scatter?

A. Rain droplets must be electrically charged
B. Rain droplets must be within the E layer
C. The rain must be within radio range of both stations
D. All of these choices are correct

C   Rain can reflect radio waves of a wide frequency range so if the rain is within radio range of both stations, it is likely they can make contact by reflecting their signals off the rain.

E3A07   Atmospheric ducts capable of propagating microwave signals often form over what geographic feature?

A. Mountain ranges
B. Forests
C. Bodies of water
D. Urban areas

C   Over bodies of water, air often forms stable layers that differ in temperature and humidity. The layers can act as guides to VHF and higher-frequency radio waves over long distances.

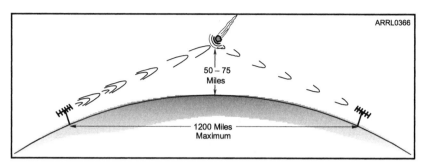

Figure E3A08 — Meteor-scatter communication makes extended-range VHF contacts possible by using the ionized meteor trail as a reflector 50-75 miles above the Earth.

E3A08 When a meteor strikes the Earth's atmosphere, a cylindrical region of free electrons is formed at what layer of the ionosphere?

    A. The E layer
    B. The F1 layer
    C. The F2 layer
    D. The D layer

A   Meteor-scatter communication makes extended-range VHF contacts possible by using the ionized meteor trail as a reflector. As a meteoroid speeds through the upper atmosphere, it begins to burn or vaporize as it collides with air molecules. This action creates heat and light and leaves a trail of free electrons and positively charged ions behind as the meteoroid races along. Trail size is directly dependent on meteoroid size and speed. A typical meteoroid is the size of a grain of sand. A particle this size creates a trail about 3 feet in diameter and 12 miles or longer, depending on speed. Meteor trails are formed at approximately the altitude of the ionospheric E layer, 50 to 75 miles above the Earth.

E3A09 Which of the following frequency ranges is most suited for meteor-scatter communications?

    A. 1.8 - 1.9 MHz
    B. 10 - 14 MHz
    C. 28 - 148 MHz
    D. 220 - 450 MHz

C   See E3A08.

E3A10   Which type of atmospheric structure can create a path for microwave propagation?

   A. The jet stream
   B. Temperature inversion
   C. Wind shear
   D. Dust devil

B   See E3A05. Temperature inversions can create stable surfaces in the atmosphere that are good guides for microwave signals which can then propagate over long distances.

E3A11   What is a typical range for tropospheric propagation of microwave signals?

   A. 10 miles to 50 miles
   B. 100 miles to 300 miles
   C. 1200 miles
   D. 2500 miles

B   The maximum range of tropospheric propagation is about the same as the size of the atmospheric structures that create the path.

E3A12   What is the cause of auroral activity?

   A. The interaction in the F2 layer between the solar wind and the Van Allen belt
   B. A low sunspot level combined with tropospheric ducting
   C. The interaction in the E layer of charged particles from the Sun with the Earth's magnetic field
   D. Meteor showers concentrated in the extreme northern and southern latitudes

C   Captured by the Earth's magnetic field, charged particles from the Sun travel along the field to the magnetic poles. Near the poles, the particles are travelling nearly vertically and as they collide with molecules in the E layer of the atmosphere, create ions that can reflect radio waves.

E3A13   Which emission mode is best for aurora propagation?

   A. CW
   B. SSB
   C. FM
   D. RTTY

A   Aurora results from a large-scale interaction between the magnetic field of the Earth and electrically charged particles arriving from the Sun. Signals received by auroral propagation are badly distorted because of the erratic nature of reflection from the auroral region. For that reason, CW is the most effective mode for auroral work. While SSB may be usable at 6 meters when signals are strong and the operator speaks slowly and distinctly, it is rarely usable at 2 meters or higher frequencies.

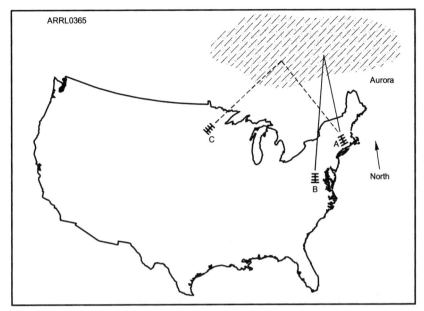

Figure E3A14 — To make contacts by reflecting signals from the aurora, stations in the Northern Hemisphere point their antennas toward the North Pole — or toward the North Magnetic Pole if they are close enough for there to be a significant difference in bearings. Aim your antenna in different directions to find strongest reflection.

E3A14   From the contiguous 48 states, in which approximate direction should an antenna be pointed to take maximum advantage of aurora propagation?

    A. South
    B. North
    C. East
    D. West

B   Auroras occur around the magnetic poles. Stations in the Northern Hemisphere point their antennas toward the North Pole — or toward the North Magnetic Pole if they are close enough for there to be a significant difference in bearings. Aim your antenna in different directions to find strongest reflection.

E3A15   What is an electromagnetic wave?

A. A wave of alternating current, in the core of an electromagnet
B. A wave consisting of two electric fields at parallel right angles to each other
C. A wave consisting of an electric field and a magnetic field oscillating at right angles to each other
D. A wave consisting of two magnetic fields at right angles to each other

C   An electromagnetic wave has both an electric and a magnetic field component. The fields are oscillating and at right angles to one another. This means, for example, when the electric field is horizontally polarized the magnetic field is vertical and vice-versa. It is the changing electric and magnetic fields that propagate electromagnetic energy across the vacuum of free space.

E3A16   Which of the following best describes electromagnetic waves traveling in free space?

A. Electric and magnetic fields become aligned as they travel
B. The energy propagates through a medium with a high refractive index
C. The waves are reflected by the ionosphere and return to their source
D. Changing electric and magnetic fields propagate the energy

D   See E3A15.

E3A17   What is meant by circularly polarized electromagnetic waves?

A. Waves with an electric field bent into a circular shape
B. Waves with a rotating electric field
C. Waves that circle the Earth
D. Waves produced by a loop antenna

B   It is possible to generate waves in which the electric and magnetic fields rotate as they travel while maintaining their right-angle orientation. This condition is called circular polarization. It is particularly helpful to use circular polarization in satellite communication, where polarization tends to shift as the signal traverses the ionosphere or the satellite moves.

E3B   Transequatorial propagation; long path; gray-line; multi-path; ordinary and extraordinary waves; chordal hop, sporadic E mechanisms

E3B01   What is transequatorial propagation?

    A. Propagation between two mid-latitude points at approximately the same distance north and south of the magnetic equator

    B. Propagation between any two points located on the magnetic equator

    C. Propagation between two continents by way of ducts along the magnetic equator

    D. Propagation between two stations at the same latitude

A   Transequatorial propagation (TE) is a form of F layer ionospheric propagation that was discovered by amateurs. TE allows hams on either side of the magnetic equator to communicate with each other. As the signal frequency increases, TE propagation becomes more restricted to regions equidistant from, and perpendicular to, the magnetic equator. The world map in Figure E3B01 shows TE paths worked by amateurs on 144 MHz.

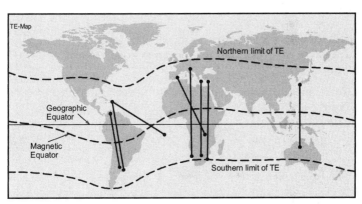

Figure E3B01 — This world map shows TE paths worked by amateurs on 144 MHz. Notice the symmetrical distribution of stations with respect to the magnetic equator. Because the poles of the Earth's magnetic field are not aligned with its geomagnetic axis, the magnetic equator does not follow the geographic equator and is somewhat tilted and distorted.

E3B02   What is the approximate maximum range for signals using transequatorial propagation?

    A. 1000 miles

    B. 2500 miles

    C. 5000 miles

    D. 7500 miles

C   See E3B01. Maximum transequatorial propagation range is approximately 5000 miles — 2500 miles on each side of the magnetic equator.

E3B03   What is the best time of day for transequatorial propagation?

A. Morning
B. Noon
C. Afternoon or early evening
D. Late at night

C   See E3B01. Ionization levels that support transequatorial propagation are reachedg during the morning, are well established by noon and may last until after midnight. The best (peak) time is in the afternoon and early evening hours.

E3B04   What is meant by the terms extraordinary and ordinary waves?

A. Extraordinary waves describe rare long skip propagation compared to ordinary waves which travel shorter distances
B. Independent waves created in the ionosphere that are elliptically polarized
C. Long path and short path waves
D. Refracted rays and reflected waves

B   A radio wave entering an ionized region like the ionosphere in which there is also a magnetic field, will split into two waves which are elliptically polarized with their E fields at right angles to each other. These are the extraordinary and ordinary waves which each take a separate path to the receiver.

E3B05   Which amateur bands typically support long-path propagation?

A. 160 to 40 meters
B. 30 to 10 meters
C. 160 to 10 meters
D. 6 meters to 2 meters

C   Long-path propagation (propagation in the opposite direction to the more-direct short path) can occur on any band that provides ionospheric propagation. That means you might experience long-path propagation on the 160 to 10 meter bands. Long-path propagation has been observed on 6 meters but it is quite uncommon.

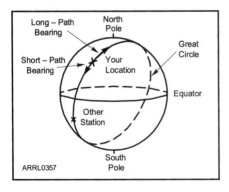

Figure E3B05 — A drawing showing a great circle path between two stations. The short-path and long-path bearings are shown from the perspective of the Northern Hemisphere station.

E3B06   Which of the following amateur bands most frequently provides long-path propagation?

A. 80 meters
B. 20 meters
C. 10 meters
D. 6 meters

B   See E3B05. You can consistently make use of long-path enhancement on the 20 meter band. All it takes is a modest beam antenna with a relatively high gain compared to a dipole, such as a Yagi or quad at a height that enhances radiation at low vertical angles.

E3B07   Which of the following could account for hearing an echo on the received signal of a distant station?

A. High D layer absorption
B. Meteor scatter
C. Transmit frequency is higher than the MUF
D. Receipt of a signal by more than one path

D   See E3B05. If you are in North America and hear an echo on signals from European stations when your antenna is pointing toward Europe, the echo may be coming in by long-path propagation. Because the signals have to travel much further on the long path, they will be delayed compared to the short-path signals.

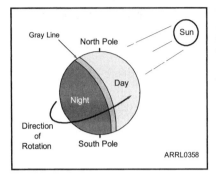

Figure E3B08 — The gray line is a transition region between daylight and darkness. One side of the Earth is coming into sunrise and the other is just past sunset.

E3B08   What type of HF propagation is probably occurring if radio signals travel along the terminator between daylight and darkness?

A. Transequatorial
B. Sporadic-E
C. Long-path
D. Gray-line

D   The gray line is a transition region along the line around the Earth between daylight and darkness. One side of the Earth is coming into sunrise and the other is just past sunset. Astronomers call this line the terminator. Propagation along the gray line can be very efficient because the D layer, which absorbs HF signals, disappears rapidly on the sunset side of the gray line, and has yet to build up on the sunrise side. By contrast, the much higher F layer forms earlier and lasts much longer.

E3B09   At what time of year is Sporadic E propagation most likely to occur?

A. Around the solstices, especially the summer solstice
B. Around the solstices, especially the winter solstice
C. Around the equinoxes, especially the spring equinox
D. Around the equinoxes, especially the fall equinox

A   The solstices, when the Earth's axis is tipped at the maximum angle with respect to the Sun, are the time of best conditions for sporadic E propagation.

E3B10   What is the cause of gray-line propagation?

A. At midday, the Sun super heats the ionosphere causing increased refraction of radio waves
B. At twilight and sunrise, D-layer absorption is low while E-layer and F-layer propagation remains high
C. In darkness, solar absorption drops greatly while atmospheric ionization remains steady
D. At mid-afternoon, the Sun heats the ionosphere decreasing radio wave refraction and the MUF

B   See E3B08.

E3B11 At what time of day is Sporadic E propagation most likely to occur?

    A. Around sunset
    B. Around sunrise
    C. Early evening
    D. Any time

D When solar and seasonal conditions are right, sporadic E propagation can take place at any time of day.

E3B12 What is the primary characteristic of chordal hop propagation?

    A. Propagation away from the great circle bearing between stations
    B. Successive ionospheric reflections without an intermediate reflection from the ground
    C. Propagation across the geomagnetic equator
    D. Signals reflected back toward the transmitting station

B Such reflections takes place at very shallow angles, allowing long distances to be covered in a minimum of reflections.

E3B13 Why is chordal hop propagation desirable?

    A. The signal experiences less loss along the path than normal skip propagation
    B. The MUF for chordal hop propagation is much lower than for normal skip propagation
    C. Atmospheric noise is lower in the direction of chordal hop propagation
    D. Signals travel faster along ionospheric chords

A See E3B12. Because ground reflections are lossy, avoiding them through successive hops in the ionosphere results in a much stronger signal at the receiving station.

E3B14 What happens to linearly polarized radio waves that split into ordinary and extraordinary waves in the ionosphere?

    A. They are bent toward the magnetic poles
    B. Their polarization is randomly modified
    C. They become elliptically polarized
    D. They become phase-locked

C Elliptical polarization means the direction of the wave's E field rotates and changes strength.

E3C Radio-path horizon; less common propagation modes; propagation prediction techniques and modeling; space weather parameters and amateur radio

E3C01 What does the term ray tracing describe in regard to radio communications?

A. The process in which an electronic display presents a pattern
B. Modeling a radio wave's path through the ionosphere
C. Determining the radiation pattern from an array of antennas
D. Evaluating high voltage sources for X-rays

B Ray tracing software evaluates the conditions along many different paths a radio wave may take between stations.

E3C02 What is indicated by a rising A or K index?

A. Increasing disruption of the geomagnetic field
B. Decreasing disruption of the geomagnetic field
C. Higher levels of solar UV radiation
D. An increase in the critical frequency

A Higher A and K indices usually indicate poorer propagation on the HF bands.

E3C03 Which of the following signal paths is most likely to experience high levels of absorption when the A index or K index is elevated?

A. Transequatorial propagation
B. Polar paths
C. Sporadic E
D. NVIS

B See E3C02. Paths that go near or through the auroral regions near the magnetic poles are the most sensitive to disruption of the geomagnetic field.

E3C04 What does the value of Bz (B sub Z) represent?

A. Geomagnetic field stability
B. Critical frequency for vertical transmissions
C. Direction and strength of the interplanetary magnetic field
D. Duration of long-delayed echoes

C B is the symbol for magnetic field strength and Z indicates the direction in which the field is measured.

E3C05 What orientation of Bz (B sub z) increases the likelihood that incoming particles from the Sun will cause disturbed conditions?

A. Southward
B. Northward
C. Eastward
D. Westward

A See E3C04. When the interplanetary and Earth's magnetic fields have the same orientation (south), it is easier for charged particles to enter the Earth's geomagnetic field, possibly causing disruptions.

E3C06 By how much does the VHF/UHF radio horizon distance exceed the geometric horizon?

A. By approximately 15 percent of the distance
B. By approximately twice the distance
C. By approximately 50 percent of the distance
D. By approximately four times the distance

A Under normal conditions, bending in the troposphere causes VHF and UHF radio waves to be returned to Earth beyond the visible horizon. The radio horizon is approximately 15% farther than the geometric horizon. Under normal conditions, the structure of the atmosphere near the Earth causes radio waves to bend into a curved path that keeps them nearer to the Earth than would be the case for true straight-line travel.

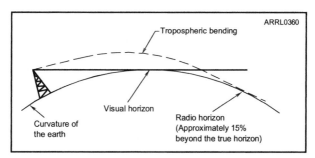

Figure E3C06 — Under normal conditions, bending in the troposphere causes VHF and UHF radio waves to be returned to Earth beyond the visible horizon.

E3C07   Which of the following descriptors indicates the greatest solar flare intensity?

    A. Class A
    B. Class B
    C. Class M
    D. Class X

D   X-class flares are the strongest, followed by M, C, and B.

E3C08   What does the space weather term G5 mean?

    A. An extreme geomagnetic storm
    B. Very low solar activity
    C. Moderate solar wind
    D. Waning sunspot numbers

A   G stands for geomagnetic field and the number indicates disruption on a scale of 1 to 5.

E3C09   How does the intensity of an X3 flare compare to that of an X2 flare?

    A. 10 percent greater
    B. 50 percent greater
    C. Twice as great
    D. Four times as great

C   See E3C07. Each increase in the numeric part of the flare's strength code corresponds to a doubling of the flare's intensity.

E3C10   What does the 304A solar parameter measure?

    A. The ratio of X-ray flux to radio flux, correlated to sunspot number
    B. The UV emission at 304 angstroms, correlated to solar flux index
    C. The solar wind velocity at 304 degrees from the solar equator, correlated to solar activity
    D. The solar emission at 304 GHz, correlated to X-Ray flare levels

B   The name of this solar parameter indicates the wavelength in angstroms at which the measurement was taken. 304 angstrom wavelength light is in the ultraviolet spectrum.

E3C11   What does VOACAP software model?

    A. AC voltage and impedance
    B. VHF radio propagation
    C. HF propagation
    D. AC current and impedance

C   VOACAP stands for Voice of America Coverage Analysis Program for HF Propagation Prediction and Ionospheric Communications Analysis. It is one of the best HF propagation prediction programs.

E3C12   How does the maximum distance of ground-wave propagation change when the signal frequency is increased?

A. It stays the same
B. It increases
C. It decreases
D. It peaks at roughly 14 MHz

C   Ground-wave propagation refers to diffraction of vertically polarized waves. Ground-wave propagation is most noticeable on the AM broadcast band and the 160 and 80 meter amateur bands. Practical ground-wave communications distances on these bands often extend to 120 miles or more. Ground-wave loss increases significantly with higher frequencies so it is not useful even at 40 meters. Although the term ground-wave propagation is often applied to any short-distance communication, the actual mechanism is unique to the lower frequencies.

E3C13   What type of polarization is best for ground-wave propagation?

A. Vertical
B. Horizontal
C. Circular
D. Elliptical

A   See E3C12. All ground-wave propagation uses vertical polarization. Signals with horizontal polarization do not propagate by ground-wave.

E3C14   Why does the radio-path horizon distance exceed the geometric horizon?

A. E-region skip
B. D-region skip
C. Downward bending due to aurora refraction
D. Downward bending due to density variations in the atmosphere

D   See E3C06. Under normal conditions, the structure of the atmosphere near the Earth causes radio waves to bend into a curved path. That effect keeps the radio waves nearer to the Earth than true straight-line travel would.

E3C15   What might a sudden rise in radio background noise indicate?

A. A meteor ping
B. A solar flare has occurred
C. Increased transequatorial propagation likely
D. Long-path propagation is occurring

B   The increased solar UV and X-rays cause the upper atmosphere to become noisier.

Amateur Practices

There will be five questions on your Extra Class examination from the Amateur Practices subelement. These five questions will be taken from the five groups of questions labeled E4A through E4E.

E4A   Test equipment: analog and digital instruments; spectrum and network analyzers, antenna analyzers; oscilloscopes; RF measurements; computer aided measurements

E4A01   Which of the following parameters determines the bandwidth of a digital or computer-based oscilloscope?

A. Input capacitance
B. Input impedance
C. Sampling rate
D. Sample resolution

C   The highest frequency signal a digital oscilloscope can display without creating aliases (false signals) is one-half of the instrument's sample rate at which it digitizes the signal. Aliases are typical signals that appear to be similar to the signal being digitized but are displayed with lower frequencies.

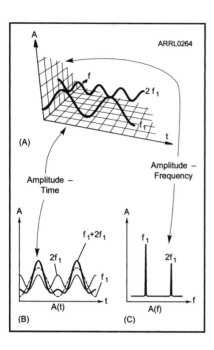

Figure E4A02 — This diagram shows how the complex signal at A can be viewed in the time domain by an oscilloscope or in the frequency domain by a spectrum analyzer. The oscilloscope would show the sum of the two signals as at B. The spectrum analyzer would show two separate components at different frequencies as at C. The spectrum analyzer is usually more useful in analyzing complex signals.

E4A02 Which of the following parameters would a spectrum analyzer display on the vertical and horizontal axes?

A. RF amplitude and time
B. RF amplitude and frequency
C. SWR and frequency
D. SWR and time

B Use a spectrum analyzer to view signals in the frequency domain (amplitude vs frequency) and an oscilloscope to view them in the time domain (amplitude vs time). For both instruments the vertical axis is signal amplitude. The difference is that the spectrum analyzer displays frequency along the horizontal axis and the oscilloscope displays time. Figure E4A02 shows how the complex signal at A can be viewed in the time domain by an oscilloscope or in the frequency domain by a spectrum analyzer. The oscilloscope would show the sum of the two signals as at B. The spectrum analyzer would show two separate components at different frequencies as at C. The spectrum analyzer is usually more useful in analyzing complex signals.

E4A03   Which of the following test instruments is used to display spurious signals and/or intermodulation distortion products in an SSB transmitter?

A. A wattmeter
B. A spectrum analyzer
C. A logic analyzer
D. A time-domain reflectometer

B   See E4A02.

E4A04   What determines the upper frequency limit for a computer soundcard-based oscilloscope program?

A. Analog-to-digital conversion speed of the soundcard
B. Amount of memory on the soundcard
C. Q of the interface of the interface circuit
D. All of these choices are correct

A   See E4A01.

E4A05   What might be an advantage of a digital vs an analog oscilloscope?

A. Automatic amplitude and frequency numerical readout
B. Storage of traces for future reference
C. Manipulation of time base after trace capture
D. All of these choices are correct

D   These features make it possible for the operator to make precise measurements that are unavailable on analog oscilloscopes.

E4A06   What is the effect of aliasing in a digital or computer-based oscilloscope?

A. False signals are displayed
B. All signals will have a DC offset
C. Calibration of the vertical scale is no longer valid
D. False triggering occurs

A   See E4A01.

E4A07 Which of the following is an advantage of using an antenna analyzer compared to an SWR bridge to measure antenna SWR?

    A. Antenna analyzers automatically tune your antenna for resonance
    B. Antenna analyzers do not need an external RF source
    C. Antenna analyzers display a time-varying representation of the modulation envelope
    D. All of these choices are correct

B   An antenna analyzer has a variable frequency low-power source built-in so keying a transmitter is not required. Antenna analyzers also provide information about the antenna's feed point impedance that is unavailable from an SWR bridge.

E4A08 Which of the following instruments would be best for measuring the SWR of a beam antenna?

    A. A spectrum analyzer
    B. A Q meter
    C. An ohmmeter
    D. An antenna analyzer

D   See E4A07.

E4A09 When using a computer's soundcard input to digitize signals, what is the highest frequency signal that can be digitized without aliasing?

    A. The same as the sample rate
    B. One-half the sample rate
    C. One-tenth the sample rate
    D. It depends on how the data is stored internally

B   See E4A01.

E4A10 Which of the following displays multiple digital signal states simultaneously?

    A. Network analyzer
    B. Bit error rate tester
    C. Modulation monitor
    D. Logic analyzer

D   A logic analyzer is similar to an oscilloscope but that has been designed to display multiple digital signals as separate traces. In addition, the logic analyzer can use the information in the signals to control the display and make measurements.

E4A11 Which of the following is good practice when using an oscilloscope probe?

A. Keep the signal ground connection of the probe as short as possible
B. Never use a high impedance probe to measure a low impedance circuit
C. Never use a DC-coupled probe to measure an AC circuit
D. All of these choices are correct

A High-frequency signals being measured with an oscilloscope probe must flow through the probe tip, to the 'scope's input circuit through a short coaxial cable, and back to the circuit through the ground connection lead. The probe tip connection and coaxial cable have very good high-frequency characteristics, but an excessively long ground connection introduces inductance into the signal path that can cause distortion or high-frequency rolloff of the signal.

E4A12 Which of the following procedures is an important precaution to follow when connecting a spectrum analyzer to a transmitter output?

A. Use high quality double shielded coaxial cables to reduce signal losses
B. Attenuate the transmitter output going to the spectrum analyzer
C. Match the antenna to the load
D. All of these choices are correct

B Most spectrum analyzer inputs are at risk of damage from input signals stronger than 1 watt. An attenuator should be used to reduce the signal amplitude coming from a transmitter before it is applied to the spectrum analyzer input.

E4A13 How is the compensation of an oscilloscope probe typically adjusted?

A. A square wave is displayed and the probe is adjusted until the horizontal portions of the displayed wave are as nearly flat as possible
B. A high frequency sine wave is displayed and the probe is adjusted for maximum amplitude
C. A frequency standard is displayed and the probe is adjusted until the deflection time is accurate
D. A DC voltage standard is displayed and the probe is adjusted until the displayed voltage is accurate

A Most oscilloscopes include a circuit that outputs a square wave calibrator signal of known voltage and frequency. The probe can be connected to the calibrator output and adjusted until the flat portions of the square wave are parallel to the zero-voltage horizontal axis, not tilted or rounded in any way. This means the probe's frequency response has been adjusted properly and input signals will not be distorted.

E4A14   What is the purpose of the prescaler function on a frequency counter?

A. It amplifies low level signals for more accurate counting
B. It multiplies a higher frequency signal so a low-frequency counter can display the operating frequency
C. It prevents oscillation in a low-frequency counter circuit
D. It divides a higher frequency signal so a low-frequency counter can display the input frequency

D   A prescaler is a frequency divider that allows a slower device to measure a high-frequency signal by re-scaling the signal's frequency, such as dividing it by 10.

E4A15   What is an advantage of a period-measuring frequency counter over a direct-count type?

A. It can run on battery power for remote measurements
B. It does not require an expensive high-precision time base
C. It provides improved resolution of low-frequency signals within a comparable time period
D. It can directly measure the modulation index of an FM transmitter

C   For very low frequency signals, instead of counting many pulses and then averaging over the gate period it is more convenient and often just as accurate to measure the period of the signal and then invert it to get frequency.

E4B Measurement technique and limitations: instrument accuracy and performance limitations; probes; techniques to minimize errors; measurement of "Q"; instrument calibration; S parameters; vector network analyzers

E4B01 Which of the following factors most affects the accuracy of a frequency counter?

    A. Input attenuator accuracy
    B. Time base accuracy
    C. Decade divider accuracy
    D. Temperature coefficient of the logic

B   A frequency counter counts the number of pulses applied to its input during a period of time and displays the results. The time base determines the period for counting pulses, so the accuracy of the time base determines the accuracy of the counter and the stability of the time base determines the stability of the frequency counter. The block diagram of the basic parts of a frequency counter. The time base is based on the output of a highly stable and accurate crystal oscillator.

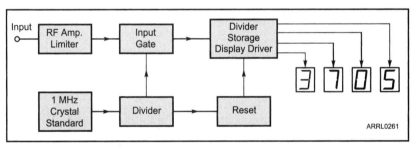

Figure E4B01 — The block diagram of the basic parts of a frequency counter. The time base is based on the output of a highly stable and accurate crystal oscillator.

E4B02   What is an advantage of using a bridge circuit to measure impedance?

    A. It provides an excellent match under all conditions
    B. It is relatively immune to drift in the signal generator source
    C. It is very precise in obtaining a signal null
    D. It can display results directly in Smith chart format

C   A bridge circuit's null is very sharp so that it is easy to adjust the bridge precisely to the deepest point of the null. That means the unknown impedance value is also measured precisely. Figure E4B03 shows a Wheatstone bridge circuit. A bridge circuit is actually a pair of voltage dividers (A) with R1 = R2. One of the dividers is adjustable as in (B). When the unknown impedance is attached at RX and the bridge adjusted so that the voltages E1 and E2 are equal, the voltmeter V indicates a null or zero voltage. The value of RS is then equal to RX.

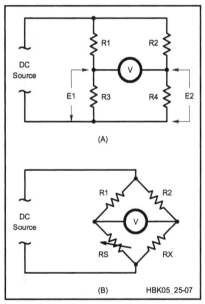

Figure E4B02 — A Wheatstone bridge circuit. A bridge circuit is actually a pair of voltage dividers (A) with R1 = R2. One of the dividers is adjustable as in (B). When the unknown impedance is attached at RX and the bridge adjusted so that the voltages E1 and E2 are equal, the voltmeter V indicates a null or zero voltage. The value of RS is then equal to RX.

E4B03   If a frequency counter with a specified accuracy of +/- 1.0 ppm reads 146,520,000 Hz, what is the most the actual frequency being measured could differ from the reading?

   A. 165.2 Hz
   B. 14.652 kHz
   C. 146.52 Hz
   D. 1.4652 MHz

C   This is a series of questions for which the answer is computed by a process based on the formula

Error = Frequency × Accuracy

In this case, when you substitute the numbers, you get

$$\text{Error} = 146,520,000 \times \frac{\pm 1}{1,000,000} = \pm 146.520 \, \text{Hz}$$

It makes the math a bit easier to realize that if you write the frequency in MHz then it will cancel with the accuracy in ppm.

E4B04   If a frequency counter with a specified accuracy of +/- 0.1 ppm reads 146,520,000 Hz, what is the most the actual frequency being measured could differ from the reading?

   A. 14.652 Hz
   B. 0.1 MHz
   C. 1.4652 Hz
   D. 1.4652 kHz

A   See E4B03.

$$\text{Error} = 146,520,000 \times \frac{\pm 0.1}{1,000,000} = \pm 14.652 \, \text{Hz}$$

E4B05   If a frequency counter with a specified accuracy of +/- 10 ppm reads 146,520,000 Hz, what is the most the actual frequency being measured could differ from the reading?

   A. 146.52 Hz
   B. 10 Hz
   C. 146.52 kHz
   D. 1465.20 Hz

D   See E4B03.

$$\text{Error} = 146,520,000 \times \frac{\pm 10}{1,000,000} = \pm 1465.20 \, \text{Hz}$$

E4B06 How much power is being absorbed by the load when a directional power meter connected between a transmitter and a terminating load reads 100 watts forward power and 25 watts reflected power?

A. 100 watts
B. 125 watts
C. 25 watts
D. 75 watts

D A directional wattmeter reads the total amount of power in a feed line traveling in each direction. The actual amount of power being absorbed by the load is the difference between the forward and reflected power readings, Forward power – Reflected power = 100 – 25 = 75 watts.

E4B07 What do the subscripts of S parameters represent?

A. The port or ports at which measurements are made
B. The relative time between measurements
C. Relative quality of the data
D. Frequency order of the measurements

A S or scattering parameters measure how much signal is reflected by a circuit (the connection is called a port) or how much is transmitted through the circuit between two ports. The numbers indicate the port at which the signal was measured and at which port the signal was applied. For example, S_{21} measures the ratio of output signal at port 2 to input signal at port 1, which is gain.

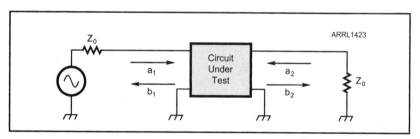

Figure E4B07 — A two-port circuit viewed as being driven by voltage waves. The relationship of the voltage waves, such as gain or SWR, is described by scattering or S parameters.

E4B08 Which of the following is a characteristic of a good DC voltmeter?

A. High reluctance input
B. Low reluctance input
C. High impedance input
D. Low impedance input

C A good voltmeter, either ac or dc, measures voltage while drawing as little current as possible from the circuit being measured. This requires the meter to have a high impedance input circuit.

E4B09 What is indicated if the current reading on an RF ammeter placed in series with the antenna feed line of a transmitter increases as the transmitter is tuned to resonance?

A. There is possibly a short to ground in the feed line
B. The transmitter is not properly neutralized
C. There is an impedance mismatch between the antenna and feed line
D. There is more power going into the antenna

D Higher feed line current means that more power is flowing to the antenna.

E4B10 Which of the following describes a method to measure intermodulation distortion in an SSB transmitter?

A. Modulate the transmitter with two non-harmonically related radio frequencies and observe the RF output with a spectrum analyzer
B. Modulate the transmitter with two non-harmonically related audio frequencies and observe the RF output with a spectrum analyzer
C. Modulate the transmitter with two harmonically related audio frequencies and observe the RF output with a peak reading wattmeter
D. Modulate the transmitter with two harmonically related audio frequencies and observe the RF output with a logic analyzer

B A spectrum analyzer is the best instrument to display and measure spurious emissions because it displays the signal frequency as well as amplitude. The test tones into the transmitter should not be harmonically related (integer multiples of the same frequency) so that any spurious signals do not occur at the same frequency. Spurious outputs from non-harmonically related signals will appear as independent signal components. See E4A02.

E4B11   How should an antenna analyzer be connected when measuring antenna resonance and feed point impedance?

    A. Loosely couple the analyzer near the antenna base

    B. Connect the analyzer via a high-impedance transformer to the antenna

    C. Loosely couple the antenna and a dummy load to the analyzer

    D. Connect the antenna feed line directly to the analyzer's connector

D   The analyzer is actually a variable-frequency low-power transmitter with a built-in impedance bridge. So SWR is measured just as with standalone transmitters and SWR meters — the feed line is connected directly to the analyzer's output.

E4B12   What is the significance of voltmeter sensitivity expressed in ohms per volt?

    A. The full scale reading of the voltmeter multiplied by its ohms per volt rating will indicate the input impedance of the voltmeter

    B. When used as a galvanometer, the reading in volts multiplied by the ohms per volt will determine the power drawn by the device under test

    C. When used as an ohmmeter, the reading in ohms divided by the ohms per volt will determine the voltage applied to the circuit

    D. When used as an ammeter, the full scale reading in amps divided by ohms per volt rating will determine the size of shunt needed

A   Ohms/volt (ohms per volt) is a measure of how sensitive the meter is because it is the reciprocal of the current (volts divided by ohms) required for a full-scale reading. Higher values of ohms/volt mean lower input currents for an equivalent voltage measurement and less loading of the circuit being tested.

E4B13   Which S parameter is equivalent to forward gain?

    A. S11

    B. S12

    C. S21

    D. S22

C   See E4B07.

E4B14   What happens if a dip meter is too tightly coupled to a tuned circuit being checked?

    A. Harmonics are generated

    B. A less accurate reading results

    C. Cross modulation occurs

    D. Intermodulation distortion occurs

B   When a dip meter it is too tightly coupled with the tuned circuit being checked, a less accurate reading results. Whenever two circuits are coupled, no matter how loosely, each circuit affects the other to some extent. Coupling that is too tight will almost certainly result in an inaccurate reading on the dip meter.

E4B15   Which of the following can be used as a relative measurement of the Q for a series-tuned circuit?

A. The inductance to capacitance ratio
B. The frequency shift
C. The bandwidth of the circuit's frequency response
D. The resonant frequency of the circuit

C   The Q of a resonant circuit equals resonant frequency (f_0) divided by bandwidth (BW). By measuring the circuit's frequency response around its resonant frequency, both f_0 and BW can be determined and Q calculated.

E4B16   Which S parameter represents return loss or SWR?

A. S11
B. S12
C. S21
D. S22

A   See E4B07. S11 measures the ratio of the reflected signal at port 1 to the input signal at port 1, from which SWR can be calculated.

E4B17   What three test loads are used to calibrate a standard RF vector network analyzer?

A. 50 ohms, 75 ohms, and 90 ohms
B. Short circuit, open circuit, and 50 ohms
C. Short circuit, open circuit, and resonant circuit
D. 50 ohms through 1/8 wavelength, 1/4 wavelength, and 1/2 wavelength of coaxial cable

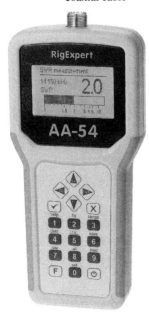

B   Similar to antenna analyzers, the vector network analyzer or VNA measures both amplitude and phase of signals at a circuit's input and output ports. The VNA can measure S parameters very easily (see E4B07). To calibrate a VNA, three test loads are used: a short-circuit (0 ohms), the system standard impedance (usually 50 ohms), and an open circuit (infinite impedance.) The results of the calibration are used to take into account the effects of the connections to the circuit.

Figure E4B17 — An antenna analyzer, such as this RigExpert AA54, consists of a tunable signal source, a frequency counter, an impedance bridge, and displays — all microprocessor controlled. The analyzer displays impedance, SWR, and frequency.

E4C Receiver performance characteristics, phase noise, noise floor, image rejection, MDS, signal-to-noise-ratio; selectivity; effects of SDR receiver non-linearity

E4C01 What is an effect of excessive phase noise in the local oscillator section of a receiver?

A. It limits the receiver's ability to receive strong signals
B. It reduces receiver sensitivity
C. It decreases receiver third-order intermodulation distortion dynamic range
D. It can cause strong signals on nearby frequencies to interfere with reception of weak signals

D One result of receiver phase noise is that as you tune closer to a strong signal, the receiver noise floor appears to increase. In other words, you hear an increasing amount of noise in an otherwise quiet receiver as you tune toward the strong signal. This reciprocal mixing means that strong signals may interfere with the reception of a nearby weak signal.

E4C02 Which of the following portions of a receiver can be effective in eliminating image signal interference?

A. A front-end filter or pre-selector
B. A narrow IF filter
C. A notch filter
D. A properly adjusted product detector

A Front-end filters and pre-selectors attenuate out-of-band signals on image frequencies so that images in the receiver are greatly reduced or eliminated. An IF or notch filter cannot prevent images because the images are produced before the IF stages. A product detector operates on whatever signal the IF stages produce, whether a real or image signal.

E4C03 What is the term for the blocking of one FM phone signal by another, stronger FM phone signal?

A. Desensitization
B. Cross-modulation interference
C. Capture effect
D. Frequency discrimination

C The capture effect in FM receivers results in the loudest signal received being the only signal demodulated, even if it is only two or three times (3 to 5 dB) stronger than other signals on the same frequency. This can be an advantage if you want to suppress interference to the stronger signal. However, the capture effect can prevent you from hearing a weaker signal in the presence of a stronger one.

E4C04 How is the noise figure of a receiver defined?

A. The ratio of atmospheric noise to phase noise
B. The ratio of the noise bandwidth in hertz to the theoretical bandwidth of a resistive network
C. The ratio of thermal noise to atmospheric noise
D. The ratio in dB of the noise generated by the receiver compared to the theoretical minimum noise

D   The noise figure of a receiver measures the noise contributed by the receiver circuits. The lower a receiver's noise figure, the lower its minimum detectible signal (MDS). If noise figure is reduced, the receiver's noise contribution is reduced. That improves (increases) the signal-to-noise ratio. Noise figure is measured in dB.

E4C05 What does a value of -174 dBm/Hz represent with regard to the noise floor of a receiver?

A. The minimum detectable signal as a function of receive frequency
B. The theoretical noise at the input of a perfect receiver at room temperature
C. The noise figure of a 1 Hz bandwidth receiver
D. The galactic noise contribution to minimum detectable signal

B   Noise in a receiver is primarily caused by temperature-related movement of electrons in the receiver's input circuits. The amount of noise power also depends on the receiver's bandwidth, with wider bandwidths allowing more noise power to be received. The theoretical noise floor limit is a level of –174 dBm in a bandwidth of 1 Hz.

E4C06 A CW receiver with the AGC off has an equivalent input noise power density of -174 dBm/Hz. What would be the level of an unmodulated carrier input to this receiver that would yield an audio output SNR of 0 dB in a 400 Hz noise bandwidth?

A. –174 dBm
B. –164 dBm
C. –155 dBm
D. –148 dBm

D   Since noise power is directly proportional to bandwidth, the ratio of filter bandwidths also determines the amount of additional noise power received according to the following formula:

$$MDS = -174 + 10 \log \left(\frac{\text{filter bandwidth}}{1\,\text{Hz}} \right)$$

Using the numbers in the question,

$$MDS = -174\,\text{dBm} + 10 \log (400) = -174\,\text{dBm} + 26\,\text{dB} = -148\,\text{dBm}$$

E4C07   What does the MDS of a receiver represent?

A. The meter display sensitivity
B. The minimum discernible signal
C. The multiplex distortion stability
D. The maximum detectable spectrum

B   The MDS or minimum discernible signal (also minimum detectable signal) is the signal level that is equal to the receiver noise floor in a specified bandwidth.

E4C08   An SDR receiver is overloaded when input signals exceed what level?

A. One-half the maximum sample rate
B. One-half the maximum sampling buffer size
C. The maximum count value of the analog-to-digital converter
D. The reference voltage of the sanalog-to-digital converter

C   Overload of a software defined radio (SDR) is reached when the combination of all signals at the receiver's analog-to-digital converter exceeds the maximum level for which the converter can generate a digital value.

E4C09   Which of the following choices is a good reason for selecting a high frequency for the design of the IF in a conventional HF or VHF communications receiver?

A. Fewer components in the receiver
B. Reduced drift
C. Easier for front-end circuitry to eliminate image responses
D. Improved receiver noise figure

C   Since image frequencies are separated from the intended receive frequency by the IF, a higher IF results in the image frequencies being farther from the intended receive frequency. That makes it easier to filter out signals at the image frequencies without affecting the desired signals.

E4C10   Which of the following is a desirable amount of selectivity for an amateur RTTY HF receiver?

A. 100 Hz
B. 300 Hz
C. 6000 Hz
D. 2400 Hz

B   The filter bandwidth should be wide enough to pass the mark and space frequencies along with their respective sidebands. These are spaced less than 200 Hz apart in the usual RTTY signal. You'll want a little bit extra bandwidth to allow for tuning error. Don't make the filter too wide or it will pass additional noise and interference.

E4C11 Which of the following is a desirable amount of selectivity for an amateur SSB phone receiver?

A. 1 kHz
B. 2.4 kHz
C. 4.2 kHz
D. 4.8 kHz

B Intelligibility of a voice signal is mostly contained in the range of 300 Hz to 2700 Hz. While wider filters may increase fidelity to some degree depending on the transmitted signal, they will pass more noise and interference. Under crowded band conditions, filters with a bandwidth as low as 1.5 kHz can be useful but at the cost of greatly reduced fidelity.

E4C12 What is an undesirable effect of using too wide a filter bandwidth in the IF section of a receiver?

A. Output-offset overshoot
B. Filter ringing
C. Thermal-noise distortion
D. Undesired signals may be heard

D Some operators like to use wide filters when tuning or monitoring a band that is quiet in terms of noise and in terms of the number of transmitting stations. Few want wider filters when the band is active and many stations are transmitting. A filter that is wider than necessary allows extra signals (and noise) to pass through the IF to the detector and on to the audio output.

E4C13 How does a narrow-band roofing filter affect receiver performance?

A. It improves sensitivity by reducing front end noise
B. It improves intelligibility by using low Q circuitry to reduce ringing
C. It improves dynamic range by attenuating strong signals near the receive frequency
D. All of these choices are correct

C A roofing filter is applied to the signal path before the final filters that have bandwidths closer to that of the desired signal. The function of a roofing filter is to remove strong nearby signals that may overload the receiver circuits before being rejected by the narrower single-signal filters.

E4C14 What transmit frequency might generate an image response signal in a receiver tuned to 14.300 MHz and which uses a 455 kHz IF frequency?

    A. 13.845 MHz
    B. 14.755 MHz
    C. 14.445 MHz
    D. 15.210 MHz

D Image frequencies are located twice the IF from the desired signal. In this case, the images are located at:

14.300 MHz ± 2 × 455 kHz = 15.210 MHz and 13.390 MHz

E4C15 What is usually the primary source of noise that can be heard from an HF receiver with an antenna connected?

    A. Detector noise
    B. Induction motor noise
    C. Receiver front-end noise
    D. Atmospheric noise

D Below 20 MHz, the largest contributor of noise is external to the receiver is atmospheric noise from lightning and other forms of electrical activity.

E4C16 Which of the following is caused by missing codes in an SDR receiver's analog-to-digital converter?

    A. Distortion
    B. Overload
    C. Loss of sensitivity
    D. Excess output level

A A missing code (one the analog-to-digital converter cannot produce) causes an apparent jump in the input signal level that acts like a transient added to an analog signal. The result is distortion (albeit small) of the input signal as it is converted to digital form.

E4C17 Which of the following has the largest effect on an SDR receiver's linearity?

    A. CPU register width in bits
    B. Anti-aliasing input filter bandwidth
    C. RAM speed used for data storage
    D. Analog-to-digital converter sample width in bits

D The more bits with which the analog-to-digital converter can represent the input signal, the more precise its measurement of the input signal. This means the steps between each successive digital value are smaller and the converter more closely approaches perfect linearity.

E4D   Receiver performance characteristics: blocking dynamic range; intermodulation and cross-modulation interference; 3rd order intercept; desensitization; preselector

E4D01   What is meant by the blocking dynamic range of a receiver?

A. The difference in dB between the noise floor and the level of an incoming signal which will cause 1 dB of gain compression
B. The minimum difference in dB between the levels of two FM signals which will cause one signal to block the other
C. The difference in dB between the noise floor and the third order intercept point
D. The minimum difference in dB between two signals which produce third order intermodulation products greater than the noise floor

A   Blocking dynamic range (BDR) refers to the ability of a receiver to respond linearly to strong signals. The definition of BDR is the difference between MDS and the input signal level at which receiver gain drops (gain compression) by 1 dB. Figure E4D01 shows how the output power of the receiver for the desired signal and the output power for the second and third-order distortion products vary with changes of the input signal power. The input consists of two equal-power sine-wave signals. Higher intercept points represent better receiver IMD performance.

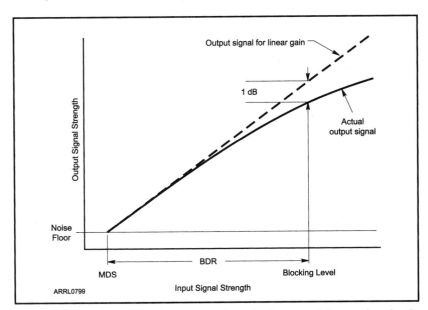

Figure E4D01 — This graph shows how the output power of the receiver for the desired signal and the output power for the second and third-order distortion products vary with changes of the input signal power. The input consists of two equal-power sine-wave signals. Higher intercept points represent better receiver IMD performance.

E4D02   Which of the following describes two problems caused by poor dynamic range in a communications receiver?

A. Cross-modulation of the desired signal and desensitization from strong adjacent signals
B. Oscillator instability requiring frequent retuning and loss of ability to recover the opposite sideband
C. Cross-modulation of the desired signal and insufficient audio power to operate the speaker
D. Oscillator instability and severe audio distortion of all but the strongest received signals

A   A receiver with poor IMD dynamic range will exhibit cross modulation of the desired signal by strong adjacent signals. One with poor blocking dynamic range will suffer from desensitization.

E4D03   How can intermodulation interference between two repeaters occur?

A. When the repeaters are in close proximity and the signals cause feedback in the final amplifier of one or both transmitters
B. When the repeaters are in close proximity and the signals mix in the final amplifier of one or both transmitters
C. When the signals from the transmitters are reflected out of phase from airplanes passing overhead
D. When the signals from the transmitters are reflected in phase from airplanes passing overhead

B   Intermodulation can be a problem in transmitters as well as receivers. This can happen when two transmitters are in close proximity and the signals mix in one or both of their final amplifiers or in a non-linear device or junction near the transmitters. This can result in severe interference.

E4D04   Which of the following may reduce or eliminate intermodulation interference in a repeater caused by another transmitter operating in close proximity?

A. A band-pass filter in the feed line between the transmitter and receiver
B. A properly terminated circulator at the output of the transmitter
C. A Class C final amplifier
D. A Class D final amplifier

B   Circulators and isolators are usually highly effective in eliminating intermodulation between two transmitters. They work like one-way valves, allowing energy to flow from the transmitter to the antenna while greatly reducing energy flow in the opposite direction. You might think that installing some type of filter would cure the problem but that's not true in this case because the offending transmitter will have a very strong signal in the filter's passband.

E4D05 What transmitter frequencies would cause an intermodulation-product signal in a receiver tuned to 146.70 MHz when a nearby station transmits on 146.52 MHz?

A. 146.34 MHz and 146.61 MHz
B. 146.88 MHz and 146.34 MHz
C. 146.10 MHz and 147.30 MHz
D. 173.35 MHz and 139.40 MHz

A The frequencies of the strongest IMD components come from the equations

$$f_{IMD} = 2f_1 \pm f_2$$

and

$$f_{IMD} = 2f_2 \pm f_1$$

You are given f_1 = 146.52 MHz and f_{IMD} = 146.70 MHz, whose sum falls into the UHF range. For that reason you'll only need to look at the differences. Solve for f_2 and you'll find

$$f_2 = 2f_1 - f_{IMD} = (2 \times 146.52 \text{ MHz}) - 146.70 \text{ MHz} = 146.34 \text{ MHz}$$

and

$$f_2 = \frac{f_1 + f_{IMD}}{2} = \frac{146.52 \text{ MHz} + 146.70 \text{ MHz}}{2} = 146.61 \text{ MHz}$$

E4D06 What is the term for unwanted signals generated by the mixing of two or more signals?

A. Amplifier desensitization
B. Neutralization
C. Adjacent channel interference
D. Intermodulation interference

D Intermodulation interference occurs when the signals of two transmitters mix together in one or both of their final amplifiers and unwanted signals at the sum and difference frequencies of the original signals are generated.

E4D07 Which describes the most significant effect of an off-frequency signal when it is causing cross-modulation interference to a desired signal?

A. A large increase in background noise
B. A reduction in apparent signal strength
C. The desired signal can no longer be heard
D. The off-frequency unwanted signal is heard in addition to the desired signal

D The term cross-modulation is used when modulation from an unwanted signal is heard in addition to the desired signal.

E4D08 What causes intermodulation in an electronic circuit?

A. Too little gain
B. Lack of neutralization
C. Nonlinear circuits or devices
D. Positive feedback

C In a linear circuit, the output is a faithful representation of the input. Nonlinearities in either circuits or devices cause distortion. This nonlinearity is the cause of intermodulation in an electronic circuit.

E4D09 What is the purpose of the preselector in a communications receiver?

A. To store often-used frequencies
B. To provide a range of AGC time constants
C. To increase rejection of unwanted signals
D. To allow selection of the optimum RF amplifier device

C Communications receivers are often operated in an environment where very strong out-of-band signals are present, such as from commercial or military stations or shortwave broadcast stations. These signals can be so strong that they overload the receiver input circuits, causing desensitization and numerous spurious signals. A preselector is a relatively broad filter that attenuates these signals, helping the receiver filter them out without being overloaded.

E4D10 What does a third-order intercept level of 40 dBm mean with respect to receiver performance?

A. Signals less than 40 dBm will not generate audible third-order intermodulation products

B. The receiver can tolerate signals up to 40 dB above the noise floor without producing third-order intermodulation products

C. A pair of 40 dBm signals will theoretically generate a third-order intermodulation product with the same level as the input signals

D. A pair of 1 mW input signals will produce a third-order intermodulation product which is 40 dB stronger than the input signal

C The third-order intercept point of a receiver is that input level where third-order IMD products equal the desired (first-order) output level. The desired output increases 1 dB for a 1-dB increase of the input. The second-order IMD increases 2 dB, and third order goes up 3 dB for each 1-dB increase in input level. A third-order intercept point of 40 dBm means that it would take a pair of signals with levels of 40 dBm (10 watts!) to create intermodulation products with the same strength as the receiver output on the desired frequency. Real-world receivers will experience blocking or gain compression before the input rises to that level and for that reason you have to compute the intercept point. It can't be measured directly.

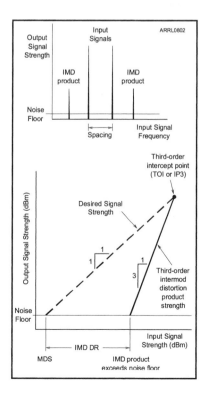

Figure E4D10 — The third-order intercept point of a receiver is that input level where third-order IMD products equal the desired (first-order) output level. Real-world receivers will experience blocking or gain compression before the input rises to that level and for that reason you have to compute the intercept point. It can't be measured directly.

E4D11   Why are third-order intermodulation products created within a receiver of particular interest compared to other products?

    A. The third-order product of two signals which are in the band of interest is also likely to be within the band
    B. The third-order intercept is much higher than other orders
    C. Third-order products are an indication of poor image rejection
    D. Third-order intermodulation produces three products for every input signal within the band of interest

A   The frequencies of the strongest IMD components come from the equations

$$f_{IMD} = 2f_1 \pm f_2$$

and

$$f_{IMD} = 2f_2 \pm f_1$$

These frequencies are close to the frequencies of the signals generating the IMD and are likely to be in the same band as the desired signal. So third-order IMD products are the most likely to cause interference.

E4D12   What is the term for the reduction in receiver sensitivity caused by a strong signal near the received frequency?

    A. Desensitization
    B. Quieting
    C. Cross-modulation interference
    D. Squelch gain rollback

A   Desensitization or desense is caused by a reduction in gain or gain compression that causes the receiver output to drop as if it were less sensitive.

E4D13   Which of the following can cause receiver desensitization?

    A. Audio gain adjusted too low
    B. Strong adjacent-channel signals
    C. Audio bias adjusted too high
    D. Squelch gain misadjusted

B   Desensitization from a strong signal near the desired signal causes the receiver circuits to be overloaded such that they can no longer amplify the desired signal by the correct amount.

E4D14 Which of the following is a way to reduce the likelihood of receiver desensitization?

A. Decrease the RF bandwidth of the receiver
B. Raise the receiver IF frequency
C. Increase the receiver front end gain
D. Switch from fast AGC to slow AGC

A Removing or attenuating the strong nearby signals is the best way of eliminating receiver desensitization. This is the function of roofing filters that remove strong signals near the desired signal before a single-signal filter determines the receiver's final selectivity.

E4E **Noise suppression: system noise; electrical appliance noise; line noise; locating noise sources; DSP noise reduction; noise blankers; grounding for signals**

E4E01 Which of the following types of receiver noise can often be reduced by use of a receiver noise blanker?

A. Ignition noise
B. Broadband white noise
C. Heterodyne interference
D. All of these choices are correct

A Noise blankers work best on impulse noise created by sharp, short pulses. Of the noise types listed, ignition noise from the firing of a vehicle's spark plugs is the best candidate for removal by a noise blanker.

E4E02 Which of the following types of receiver noise can often be reduced with a DSP noise filter?

A. Broadband white noise
B. Ignition noise
C. Power line noise
D. All of these choices are correct

D DSP noise filters can handle a wider range of noise characteristics than an analog filter. All three noise types listed here are can be removed by DSP filters.

E4E03 Which of the following signals might a receiver noise blanker be able to remove from desired signals?

A. Signals which are constant at all IF levels
B. Signals which appear across a wide bandwidth
C. Signals which appear at one IF but not another
D. Signals which have a sharply peaked frequency distribution

B Noise blankers work by detecting signals that simultaneously appear across a wide bandwidth, a characteristic of impulse noise pulses.

E4E04   How can conducted and radiated noise caused by an automobile alternator be suppressed?

    A. By installing filter capacitors in series with the DC power lead and by installing a blocking capacitor in the field lead

    B. By installing a noise suppression resistor and a blocking capacitor in both leads

    C. By installing a high-pass filter in series with the radio's power lead and a low-pass filter in parallel with the field lead

    D. By connecting the radio's power leads directly to the battery and by installing coaxial capacitors in line with the alternator leads

D   Conducted and radiated noise caused by an automobile alternator can be suppressed by connecting the radio's power leads directly to the battery which supplies the cleanest power and by installing coaxial capacitors in the alternator leads to filter out noise from that source.

E4E05   How can noise from an electric motor be suppressed?

    A. By installing a high pass filter in series with the motor's power leads

    B. By installing a brute-force AC-line filter in series with the motor leads

    C. By installing a bypass capacitor in series with the motor leads

    D. By using a ground-fault current interrupter in the circuit used to power the motor

B   A brute-force, high-pass ac line filter in series with the motor leads can suppress noise from an ac electric motor.

E4E06   What is a major cause of atmospheric static?

    A. Solar radio frequency emissions

    B. Thunderstorms

    C. Geomagnetic storms

    D. Meteor showers

B   Lightning generated in thunderstorms is a major cause of atmospheric static which can propagate over long distances. The other sources presented here do not produce large amounts of atmospheric static.

E4E07   How can you determine if line noise interference is being generated within your home?

    A. By checking the power line voltage with a time domain reflectometer

    B. By observing the AC power line waveform with an oscilloscope

    C. By turning off the AC power line main circuit breaker and listening on a battery operated radio

    D. By observing the AC power line voltage with a spectrum analyzer

C   If the line noise disappears when the ac power to the home is removed, then the noise source is supplied by the circuits in the home.

E4E08   What type of signal is picked up by electrical wiring near a radio antenna?

A. A common-mode signal at the frequency of the radio transmitter
B. An electrical-sparking signal
C. A differential-mode signal at the AC power line frequency
D. Harmonics of the AC power line frequency

A   Common-mode means that the electrical wiring is working like an antenna and the signal is being picked up by all of the conductors in common.

E4E09   What undesirable effect can occur when using an IF noise blanker?

A. Received audio in the speech range might have an echo effect
B. The audio frequency bandwidth of the received signal might be compressed
C. Nearby signals may appear to be excessively wide even if they meet emission standards
D. FM signals can no longer be demodulated

C   Because noise blankers work by detecting strong signals across a wide bandwidth, strong local signals can cause the receiver blanking circuit to activate by mistake. This causes the strong signal to sound distorted in the receiver and also causes distortion of nearby signals, making them appear to have an excessively wide bandwidth. If a strong signal seems to have excessive bandwidth, turn off the receiver noise blanker to see if that might be the problem.

E4E10   What is a common characteristic of interference caused by a touch controlled electrical device?

A. The interfering signal sounds like AC hum on an AM receiver or a carrier modulated by 60 Hz hum on a SSB or CW receiver
B. The interfering signal may drift slowly across the HF spectrum
C. The interfering signal can be several kHz in width and usually repeats at regular intervals across a HF band
D. All of these choices are correct

D   Touch-controlled devices operate by detecting a shift in an oscillator's frequency when the device is touched. With poor ac line filtering (or none at all), these oscillators often generate interference at harmonics of their operating frequency that may be modulated by the ac power 60 Hz frequency. The oscillators are also rather unstable and so the interference often drifts in frequency.

E4E11   Which is the most likely cause if you are hearing combinations of local AM broadcast signals within one or more of the MF or HF ham bands?

A. The broadcast station is transmitting an over-modulated signal
B. Nearby corroded metal joints are mixing and re-radiating the broadcast signals
C. You are receiving sky wave signals from a distant station
D. Your station receiver IF amplifier stage is defective

B   Local AM broadcast stations generate very strong signals that can cause current to flow in any metal conductor, such as fences, gutters, metal roofs and siding, pipes, and so forth. Because metal joints exposed to the weather are often corroded and nonlinear, they act as mixers, creating mixing products at various combinations of the broadcast station frequencies.

E4E12   What is one disadvantage of using some types of automatic DSP notch-filters when attempting to copy CW signals?

A. The DSP filter can remove the desired signal at the same time as it removes interfering signals
B. Any nearby signal passing through the DSP system will overwhelm the desired signal
C. Received CW signals will appear to be modulated at the DSP clock frequency
D. Ringing in the DSP filter will completely remove the spaces between the CW characters

A   A CW signal can appear much like an interfering carrier to a DSP notch filter. The result is that the filter will automatically remove the CW signal along with the properly-removed tone. For that reason, using the auto-notch filters doesn't usually work well on CW.

E4E13   What might be the cause of a loud roaring or buzzing AC line interference that comes and goes at intervals?

A. Arcing contacts in a thermostatically controlled device
B. A defective doorbell or doorbell transformer inside a nearby residence
C. A malfunctioning illuminated advertising display
D. All of these choices are correct

D   Power line noise, especially if the source is nearby, can be very loud and contain many harmonics of the 60 Hz line frequency. The resulting signal has the roaring, buzzing quality mentioned in the question. Along with sources such as defective power line insulators, line noise can also be generated by any kind of electrical contacts, switches, or lighting equipment. The arcing associated with this type of noise can be a fire hazard in a business or home.

E4E14 What is one type of electrical interference that might be caused by the operation of a nearby personal computer?

A. A loud AC hum in the audio output of your station receiver
B. A clicking noise at intervals of a few seconds
C. The appearance of unstable modulated or unmodulated signals at specific frequencies
D. A whining type noise that continually pulses off and on

C Interference from a computer (or any microprocessor-controlled device) usually consists of signals created by the sharp rise and fall of the myriad digital signals in the computer. These signals are unmodulated, so they are heard as tones. The signal frequencies change as the digital signal changes and as the computer's internal clock frequencies change with temperature.

E4E15 Which of the following can cause shielded cables to radiate or receive interference?

A. Low inductance ground connections at both ends of the shield
B. Common mode currents on the shield and conductors
C. Use of braided shielding material
D. Tying all ground connections to a common point resulting in differential mode currents in the shield

B Common-mode currents can be picked up and carried by the outer surface of shielded cables, such as coax. The currents will radiate just as any current on a wire will radiate. If the current is allowed to enter equipment at a cable's open end or an unshielded connection, it can also cause interference.

E4E16 What current flows equally on all conductors of an unshielded multi-conductor cable?

A. Differential-mode current
B. Common-mode current
C. Reactive current only
D. Return current

B Common-mode refers to current that flows equally on all conductors of a multi-conductor cable (including coax) or on the outer surface of a cable's shield.

Electrical Principles

There will be four questions on your Extra class examination from the Electrical Principles subelement. These four questions will be taken from the four groups of questions labeled E5A through E5D.

E5A Resonance and Q: characteristics of resonant circuits; series and parallel resonance; definitions and effects of Q; half-power bandwidth; phase relationships in reactive circuits

E5A01 What can cause the voltage across reactances in series to be larger than the voltage applied to them?

A. Resonance
B. Capacitance
C. Conductance
D. Resistance

A At resonance, the voltage across the inductor and capacitor in a series circuit can be many times greater than the applied voltage. In practical circuits, it can be 10 or 100 times greater. How can this be? The reason is that the capacitor and inductor store the supplied energy, dissipating a small amount of it in resistive losses. The applied voltage pumps the resonant circuit, building up energy in each component.

E5A02 What is resonance in an electrical circuit?

A. The highest frequency that will pass current
B. The lowest frequency that will pass current
C. The frequency at which the capacitive reactance equals the inductive reactance
D. The frequency at which the reactive impedance equals the resistive impedance

C At resonance, capacitive and inductive reactance in a circuit are equal.

E5A03   What is the magnitude of the impedance of a series RLC circuit at resonance?

A. High, as compared to the circuit resistance
B. Approximately equal to capacitive reactance
C. Approximately equal to inductive reactance
D. Approximately equal to circuit resistance

D   In a series circuit at resonance, the inductive reactance of L and the capacitive reactance of C cancel, leaving only the circuit resistance.

E5A04   What is the magnitude of the impedance of a circuit with a resistor, an inductor and a capacitor all in parallel, at resonance?

A. Approximately equal to circuit resistance
B. Approximately equal to inductive reactance
C. Low, as compared to the circuit resistance
D. Approximately equal to capacitive reactance

A   In a parallel circuit the basic principle of resonance is the same as for series circuits: the inductive reactance of L and the capacitive reactance of C cancel so all that remains is the circuit resistance.

E5A05   What is the magnitude of the current at the input of a series RLC circuit as the frequency goes through resonance?

A. Minimum
B. Maximum
C. R/L
D. L/R

B   Current in a series circuit is maximum at resonance because the inductive and capacitive reactances cancel.

E5A06 What is the magnitude of the circulating current within the components of a parallel LC circuit at resonance?

A. It is at a minimum
B. It is at a maximum
C. It equals 1 divided by the quantity 2 times pi, multiplied by the square root of inductance L multiplied by capacitance C
D. It equals 2 multiplied by pi, multiplied by frequency F, multiplied by inductance L

B See E5A07. The circulating current within the components of a parallel LC circuit are maximum at resonance. Circulating current transfers the circuit's stored energy back and forth between the inductive and capacitive reactance. When the two types of reactance are equal or balanced at resonance, the circulating current is at a maximum even though net current through the entire circuit is at a minimum.

E5A07 What is the magnitude of the current at the input of a parallel RLC circuit at resonance?

A. Minimum
B. Maximum
C. R/L
D. L/R

A At resonance, the circulating current of a parallel circuit is out of phase with the applied current. The effect is that very little net current flows through the resonant circuit and its impedance is very high. Current in a parallel RLC circuit does not go to zero at resonance because of the remaining resistance.

E5A08 What is the phase relationship between the current through and the voltage across a series resonant circuit at resonance?

A. The voltage leads the current by 90 degrees
B. The current leads the voltage by 90 degrees
C. The voltage and current are in phase
D. The voltage and current are 180 degrees out of phase

C At resonance, the impedance of the circuit is completely resistive because the inductive and capacitive reactances have cancelled, so the circulating current of a series circuit is in phase with the applied current. The effect is that maximum current flows through the resonant circuit and its impedance is very low.

E5A09 How is the Q of an RLC parallel resonant circuit calculated?

    A. Reactance of either the inductance or capacitance divided by the resistance

    B. Reactance of either the inductance or capacitance multiplied by the resistance

    C. Resistance divided by the reactance of either the inductance or capacitance

    D. Reactance of the inductance multiplied by the reactance of the capacitance

C   Q for series and parallel circuits calculated as follows:

$$Q_{SERIES} = \frac{X}{R} \text{ and } Q_{PARALLEL} = \frac{R}{X}$$

E5A10 How is the Q of an RLC series resonant circuit calculated?

    A. Reactance of either the inductance or capacitance divided by the resistance

    B. Reactance of either the inductance or capacitance times the resistance

    C. Resistance divided by the reactance of either the inductance or capacitance

    D. Reactance of the inductance times the reactance of the capacitance

A   See E5A09.

E5A11 What is the half-power bandwidth of a parallel resonant circuit that has a resonant frequency of 7.1 MHz and a Q of 150?

    A. 157.8 Hz

    B. 315.6 Hz

    C. 47.3 kHz

    D. 23.67 kHz

C   The relationship between bandwidth (BW), resonant frequency (f_0), and quality factor (Q) is:

$$BW = \frac{f_0}{Q} = \frac{7100 \text{ kHz}}{150} = 47.3 \text{ kHz}$$

E5A12 What is the half-power bandwidth of a parallel resonant

circuit that has a resonant frequency of 3.7 MHz and a Q of 118?

    A. 436.6 kHz
    B. 218.3 kHz
    C. 31.4 kHz
    D. 15.7 kHz

C   The relationship between bandwidth (BW), resonant frequency (f_0), and quality factor (Q) is:

$$BW = \frac{f_0}{Q} = \frac{3700 \text{ kHz}}{118} = 31.4 \text{ kHz}$$

E5A13   **What is an effect of increasing Q in a resonant circuit?**

    A. Fewer components are needed for the same performance
    B. Parasitic effects are minimized
    C. Internal voltages and circulating currents increase
    D. Phase shift can become uncontrolled

C   Tuned and impedance matching circuits work by storing energy and transferring it back and forth between the various reactive components. As Q increases, so does the amount of energy stored, causing a corresponding increase in voltages and currents in the components.

E5A14   **What is the resonant frequency of a series RLC circuit if R is 22 ohms, L is 50 microhenries and C is 40 picofarads?**

    A. 44.72 MHz
    B. 22.36 MHz
    C. 3.56 MHz
    D. 1.78 MHz

C   The resonant frequency, f_0, is the frequency at which X_L and X_C are equal. That can be shown to be

$$f_0 = \frac{1}{2\pi \sqrt{LC}}$$

The equation is the same for parallel and series resonant circuits because the only requirement is that the two reactances are equal, regardless of how they are connected together. Converting μH to H and pF to F,

$$f_0 = \frac{1}{6.28 \times \sqrt{50 \times 10^{-6} \times 40 \times 10^{-12}}} = 3,560,617 \text{ Hz} = 3.56 \text{ MHz}$$

Note that the value of R does not affect the resonant frequency of the circuit.

E5A15 Which of the following can increase Q for inductors and capacitors?

A. Lower losses
B. Lower reactance
C. Lower self-resonant frequency
D. Higher self-resonant frequency

A   Q for an inductor or capacitor is the ratio of its reactance to loss resistance, so higher Q results from lower losses.

E5A16 What is the resonant frequency of a parallel RLC circuit if R is 33 ohms, L is 50 microhenries and C is 10 picofarads?

A. 23.5 MHz
B. 23.5 kHz
C. 7.12 kHz
D. 7.12 MHz

D   See E5A14.

$$f_0 = \frac{1}{6.28 \times \sqrt{50 \times 10^{-6} \times 10 \times 10^{-12}}} = 7.12 \text{ MHz}$$

Watch out because C and D both have the same numeric answer, but C shows kHz! This is an easy one to miss if you're not careful.

E5A17 What is the result of increasing the Q of an impedance-matching circuit?

A. Matching bandwidth is decreased
B. Matching bandwidth is increased
C. Matching range is increased
D. It has no effect on impedance matching

A   As Q of an impedance-matching circuit increases, either through larger impedance transformation ratios or lower component losses, the balancing of currents and voltages creating the match occurs over a narrow frequency range. This is similar to the decreasing bandwidth of a resonant circuit as Q increases.

E5B Time constants and phase relationships: RLC time constants; definition; time constants in RL and RC circuits; phase angle between voltage and current; phase angles of series RLC; phase angle of inductance vs susceptance; admittance and susceptance

E5B01 What is the term for the time required for the capacitor in an RC circuit to be charged to 63.2% of the applied voltage?

A. An exponential rate of one
B. One time constant
C. One exponential period
D. A time factor of one

B In an RC circuit, when the capacitor has no initial charge, it takes one time constant to charge the capacitor to 63.2% of the applied voltage. In Figure E5B01, you can see how the voltage across a capacitor rises with time, when charged through a resistor. The symbol t is used to indicate a period equal to the time constant. The graphs show voltage across a capacitor as it charges (A) and discharges (B) through a resistor. The symbol τ is used for time constant and τ = R × C.

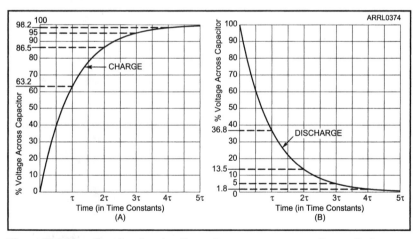

Figure E5B01 — Graphs showing the voltage across a capacitor as it charges (A) and discharges (B) through a resistor. The time it takes the capacitor to charge to 63.2% of the applied voltage or to discharge to 36.8% of an initial voltage is the circuit's time constant. The symbol τ is used for time constant.

E5B02   What is the term for the time it takes for a charged capacitor in an RC circuit to discharge to 36.8% of its initial voltage?

A. One discharge period
B. An exponential discharge rate of one
C. A discharge factor of one
D. One time constant

D   See E5B01. The voltage while charging and discharging has the same general exponential shape, so it takes one time constant to discharge 63.2% of the initial voltage, which leaves 36.8% of the initial voltage across the capacitor.

E5B03   What happens to the phase angle of a reactance when it is converted to a susceptance?

A. It is unchanged
B. The sign is reversed
C. It is shifted by 90 degrees
D. The susceptance phase angle is the inverse of the reactance phase angle

B   Similar to the relationship of conductance to resistance, susceptance (B) is the reciprocal of reactance (X). When a reactance is converted to susceptance, the magnitude of the reactance is inverted $(1/|X|)$ and the phase angle of the reactance is reversed $(-\theta)$.

E5B04   What is the time constant of a circuit having two 220-microfarad capacitors and two 1-megohm resistors, all in parallel?

A. 55 seconds
B. 110 seconds
C. 440 seconds
D. 220 seconds

D   See E5B01. In this parallel circuit, the total resistance is 500 kΩ. The total capacitance is 440 µF. The time constant is

$$\tau = R \times C = (500 \times 103) \times (440 \times 10^{-6}) = 220 \text{ seconds}$$

E5B05   What happens to the magnitude of a reactance when it is converted to a susceptance?

A. It is unchanged
B. The sign is reversed
C. It is shifted by 90 degrees
D. The magnitude of the susceptance is the reciprocal of the magnitude of the reactance

D   See E5B03.

E5B06   **What is susceptance?**

    A. The magnetic impedance of a circuit
    B. The ratio of magnetic field to electric field
    C. The inverse of reactance
    D. A measure of the efficiency of a transformer

C   See E5B03.

E5B07   **What is the phase angle between the voltage across and the current through a series RLC circuit if XC is 500 ohms, R is 1 kilohm, and XL is 250 ohms?**

    A. 68.2 degrees with the voltage leading the current
    B. 14.0 degrees with the voltage leading the current
    C. 14.0 degrees with the voltage lagging the current
    D. 68.2 degrees with the voltage lagging the current

C   You should be familiar with polar coordinates to answer this question. The total reactance in this series configuration is $250\ \Omega - 500\ \Omega = -250\ \Omega$. The phase angle between the voltage and the current is:

$$\tan^{-1}\left(\frac{X}{R}\right) = \tan^{-1}\left(\frac{-250\ \Omega}{1000\ \Omega}\right) = -14.0°$$

Because the angle is negative, the voltage lags the current. Since the net reactance is negative, the phase angle needs to be negative. Because the net reactance is smaller than the resistance, the phase angle will be less than 45°.

E5B08   **What is the phase angle between the voltage across and the current through a series RLC circuit if XC is 100 ohms, R is 100 ohms, and XL is 75 ohms?**

    A. 14 degrees with the voltage lagging the current
    B. 14 degrees with the voltage leading the current
    C. 76 degrees with the voltage leading the current
    D. 76 degrees with the voltage lagging the current

A   The total reactance in this series configuration is $75\ \Omega - 100\ \Omega$. The phase angle between the voltage and the current is

$$\tan^{-1}\left(\frac{-25\ \Omega}{100\ \Omega}\right) = -14°$$

Again, voltage lags current because the phase angle is negative.

E5B09 What is the relationship between the current through a capacitor and the voltage across a capacitor?
- A. Voltage and current are in phase
- B. Voltage and current are 180 degrees out of phase
- C. Voltage leads current by 90 degrees
- D. Current leads voltage by 90 degrees

D For a capacitor, the current leads voltage by 90°. Use the mnemonic Eli the iceman — it's an easy way to remember current and voltage relationships in reactive circuits. ELI means voltage (E) leads current (I) in an inductance (L) and ICE means current (I) leads voltage (E) in a capacitor (C). For a pure reactance (no resistance), the phase angle is 90°.

E5B10 What is the relationship between the current through an inductor and the voltage across an inductor?
- A. Voltage leads current by 90 degrees
- B. Current leads voltage by 90 degrees
- C. Voltage and current are 180 degrees out of phase
- D. Voltage and current are in phase

A See E5B09. For an inductor, the voltage leads current by 90°.

E5B11 What is the phase angle between the voltage across and the current through a series RLC circuit if XC is 25 ohms, R is 100 ohms, and XL is 50 ohms?
- A. 14 degrees with the voltage lagging the current
- B. 14 degrees with the voltage leading the current
- C. 76 degrees with the voltage lagging the current
- D. 76 degrees with the voltage leading the current

B The total reactance in this series configuration is 50 Ω – 25 Ω. The phase angle between the voltage and the current is:

$$\tan^{-1}\left(\frac{25\ \Omega}{100\ \Omega}\right) = 14°$$

The positive angle means that voltage leads current.

A rule of thumb is that since the net reactance is positive, the phase angle needs to be positive; because the net reactance is smaller than the resistance, the phase angle needs to be less than 45°.

E5B12 What is admittance?

    A. The inverse of impedance
    B. The term for the gain of a field effect transistor
    C. The turns ratio of a transformer
    D. The unit used for Q factor

A   Similar to the relationship of conductance to resistance, admittance (Y) is the reciprocal of impedance (Z). When impedance is converted to admittances, the magnitude of the admittance is inverted ($1/|Z|$) and the phase angle of the impedance is reversed ($-\theta$).

E5B13 What letter is commonly used to represent susceptance?

    A. G
    B. X
    C. Y
    D. B

D   B represents susceptance, G represents conductance, X represents reactance, Y represents admittance, and Z represents impedance.

E5C   Coordinate systems and phasors in electronics: rectangular coordinates; polar coordinates phasors

E5C01   Which of the following represents a capacitive reactance in rectangular notation?

    A. $-jX$
    B. $+jX$
    C. X
    D. Omega

A   In rectangular coordinates, impedance is the sum of the resistive (R) and reactive (X) components. While resistance is always positive, reactance can be either positive (inductive) or negative (capacitive).

E5C02   How are impedances described in polar coordinates?

    A. By X and R values
    B. By real and imaginary parts
    C. By phase angle and amplitude
    D. By Y and G values

C   Impedances in polar coordinates are represented by a distance from the origin (magnitude) and an angle from the 0-degree axis (phase angle).

E5C03   Which of the following represents an inductive reactance in polar coordinates?

  A. A positive real part
  B. A negative real part
  C. A positive phase angle
  D. A negative phase angle

C   Inductive reactance always has a positive phase angle and is plotted along the vertical, imaginary axis so it has no real part.

E5C04   Which of the following represents a capacitive reactance in polar coordinates?

  A. A positive real part
  B. A negative real part
  C. A positive phase angle
  D. A negative phase angle

D   Capacitive reactance always has a negative phase angle and is plotted along the vertical, imaginary axis so it has no real part.

E5C05   What is the name of the diagram used to show the phase relationship between impedances at a given frequency?

  A. Venn diagram
  B. Near field diagram
  C. Phasor diagram
  D. Far field diagram

C   Phasors are a special way of representing a vector that can be used for signals or for impedances as long as the frequency is the same for all signals and impedances. Phasors are very useful for calculations because the reference frequency is assumed to be the same for all of the quantities involved so you can avoid having to include frequency in the calculations.

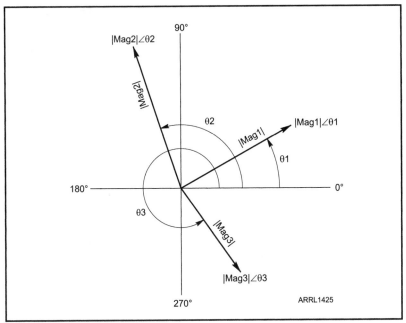

Figure E5C05 — A phasor diagram allows several phasors to be compared graphically as individual vectors. If the phasor diagram refers to signals, then the 0-degree axis represents the reference phase. If the phasor diagram refers to complex impedances, then the 0-degree axis represents the resistive component.

E5C06 What does the impedance 50 – *j*25 represent?

A. 50 ohms resistance in series with 25 ohms inductive reactance
B. 50 ohms resistance in series with 25 ohms capacitive reactance
C. 25 ohms resistance in series with 50 ohms inductive reactance
D. 25 ohms resistance in series with 50 ohms capacitive reactance

B Impedance is written as a complex number in the form R + *j*X where the magnitude of X can be positive (inductive) or negative (capacitive).

E5C07 What is a vector?

A. The value of a quantity that changes over time
B. A quantity with both magnitude and an angular component
C. The inverse of the tangent function
D. The inverse of the sine function

B Vectors combine magnitude with a directional relationship. For electrical circuits and signals, the direction is usually a phase angle.

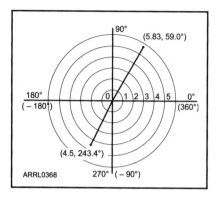

Figure E5C08 — Polar-coordinate graphs use a radius from the origin and an angle from the 0° axis to specify the location of a point. Thus, the location of any point can be specified in terms of a radius and an angle.

E5C08   What coordinate system is often used to display the phase angle of a circuit containing resistance, inductive and/or capacitive reactance?

A. Maidenhead grid
B. Faraday grid
C. Elliptical coordinates
D. Polar coordinates

D   Polar coordinate graphs consist of a radius (distance) from the origin and an angle from the right-hand horizontal axis which represents 0°. Positive angles increase in the counter-clockwise direction.

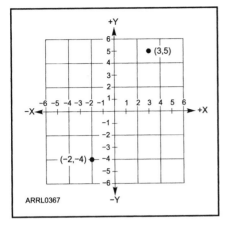

Figure E5C09 — Rectangular-coordinate graphs use a pair of axes at right angles to each other, each calibrated in numeric units. Any point on the resulting grid can be expressed in terms of its horizontal (X) and vertical (Y) values, called coordinates.

ARRL0367

E5C09 When using rectangular coordinates to graph the impedance of a circuit, what does the horizontal axis represent?

A. Resistive component
B. Reactive component
C. The sum of the reactive and resistive components
D. The difference between the resistive and reactive components

A   The horizontal axis represents the resistive component and the vertical axis represents reactance.

E5C10 When using rectangular coordinates to graph the impedance of a circuit, what does the vertical axis represent?

A. Resistive component
B. Reactive component
C. The sum of the reactive and resistive components
D. The difference between the resistive and reactive components

B   See E5C09.

E5C11 What do the two numbers that are used to define a point on a graph using rectangular coordinates represent?

A. The magnitude and phase of the point
B. The sine and cosine values
C. The coordinate values along the horizontal and vertical axes
D. The tangent and cotangent values

C   See E5C09. The numbers given to graph a point in rectangular coordinates represent values along the horizontal and vertical axes.

E5C12 If you plot the impedance of a circuit using the rectangular coordinate system and find the impedance point falls on the right side of the graph on the horizontal axis, what do you know about the circuit?

A. It has to be a direct current circuit
B. It contains resistance and capacitive reactance
C. It contains resistance and inductive reactance
D. It is equivalent to a pure resistance

D　See E5C09. If the point is on the horizontal axis (resistance), it has no reactive (Y-axis) component. That means you have a pure resistance.

E5C13 What coordinate system is often used to display the resistive, inductive, and/or capacitive reactance components of an impedance?

A. Maidenhead grid
B. Faraday grid
C. Elliptical coordinates
D. Rectangular coordinates

D　See E5C09.

E5C14 Which point on Figure E5-2 best represents that impedance of a series circuit consisting of a 400 ohm resistor and a 38 picofarad capacitor at 14 MHz?

A. Point 2
B. Point 4
C. Point 5
D. Point 6

B　Using the equation $Z = R + jX$, the total impedance of this circuit is

$$Z = R + X = R + \frac{1}{j\,2\pi\,f\,C}$$

Using the values in the question and remembering that $1/j = -j$,

$$Z = 400 + \frac{1}{j\,6.28 \times (14 \times 10^6) \times (38 \times 10^{-12})} = 400 - j300\ \Omega$$

Find the point by moving horizontally to +400 then vertically (down) to –400 (remember, capacitive reactance is negative), which corresponds to Point 4.

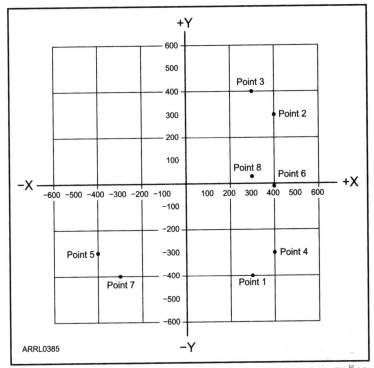

Figure E5-2 — This figure is used for questions E5C14, E5C15, E5C16, and E5C17

E5C15 Which point in Figure E5-2 best represents the impedance of a series circuit consisting of a 300 ohm resistor and an 18 microhenry inductor at 3.505 MHz?

    A. Point 1
    B. Point 3
    C. Point 7
    D. Point 8

B   Using the values in the question,

$$Z = 300 + j6.28 \times (3.505 \times 10^6) \times (18 \times 10^{-6}) = 300 + j400 \ \Omega$$

To find this point, find +300 on the horizontal axis, and then move up (positive direction) to +400 on the vertical axis. This corresponds to Point 3.

E5C16 Which point on Figure E5-2 best represents the impedance of a series circuit consisting of a 300 ohm resistor and a 19 picofarad capacitor at 21.200 MHz?

    A. Point 1
    B. Point 3
    C. Point 7
    D. Point 8

A   Using the equation Z = R + jX, the total impedance of this circuit is

$$Z = 300 + \frac{1}{j6.28 \times (21.2 \times 10^6) \times (19 \times 10^{-12})} = 300 - j400 \ \Omega$$

Find the point by moving horizontally to +300 then vertically (down) to –400 (remember, capacitive reactance is negative), which corresponds to Point 1.

E5C17 Which point on Figure E5-2 best represents the impedance of a series circuit consisting of a 300-ohm resistor, a 0.64-microhenry inductor and an 85-picofarad capacitor at 24.900 MHz?

    A. Point 1
    B. Point 3
    C. Point 5
    D. Point 8

D   The total impedance of this circuit is

$$Z = R + j\,(X_L - X_C)$$

$$Z = 300 + j\left(2\pi f L - \frac{1}{2\pi f C}\right)$$

$$Z = 300 + j\left([6.28 \times (24.9 \times 10^6) \times (0.64 \times 10^{-6})] - \frac{1}{6.28 \times (24.9 \times 10^6) \times (85 \times 10^{-12})}\right)$$

$$Z = 300 + j\,(100 - 75) = 300 + j\,25 \ \Omega$$

Find the point by moving horizontally to +300 then vertically (up) to 25, which corresponds to Point 8

E5D AC and RF energy in real circuits: skin effect; electrostatic and electromagnetic fields; reactive power; power factor; electrical length of conductors at UHF and microwave frequencies

E5D01   What is the result of skin effect?

A. As frequency increases, RF current flows in a thinner layer of the conductor, closer to the surface
B. As frequency decreases, RF current flows in a thinner layer of the conductor, closer to the surface
C. Thermal effects on the surface of the conductor increase the impedance
D. Thermal effects on the surface of the conductor decrease the impedance

A   Skin effect occurs at RF. As the frequency increases, electric and magnetic fields of the signal don't penetrate as deeply into a conductor. This results in RF current flowing in a progressively thinner layer near the surface of the conductor as the frequency increases.

E5D02   Why is it important to keep lead lengths short for components used in circuits for VHF and above?

A. To increase the thermal time constant
B. To avoid unwanted inductive reactance
C. To maintain component lifetime
D. All of these choices are correct

B   Even straight wires have inductance. At VHF and higher frequencies, the inductive reactance of component leads becomes increasingly significant so it important to keep leads short to keep them from affecting circuit behavior.

E5D03   What is microstrip?

A. Lightweight transmission line made of common zip cord
B. Miniature coax used for low power applications
C. Short lengths of coax mounted on printed circuit boards to minimize time delay between microwave circuits
D. Precision printed circuit conductors above a ground plane to provide constant impedance interconnects at microwave frequencies

D   Microstrip is type of printed-circuit board transmission line consisting of a signal trace above or between ground plane layers.

E5D04 Why are short connections necessary at microwave frequencies?

A. To increase neutralizing resistance
B. To reduce phase shift along the connection
C. Because of ground reflections
D. To reduce noise figure

B See E5D02. The longer the lead, the longer it is in terms of wavelength and the amount of phase shift in a signal traveling along the lead.

E5D05 Which parasitic characteristic increases with conductor length?

A. Inductance
B. Permeability
C. Permittivity
D. Malleability

A See E5D02.

E5D06 In what direction is the magnetic field oriented about a conductor in relation to the direction of electron flow?

A. In the same direction as the current
B. In a direction opposite to the current
C. In all directions; omni-directional
D. In a direction determined by the left-hand rule

D A magnetic field curls around electrical current. The direction of a magnetic field around a conductor can be determined by using the left-hand rule. With the wire going across the palm of your left hand, and with your thumb pointed in the direction of electron flow, wrap your fingers around the wire. Your fingers will be pointing in the direction of the magnetic field. This is called the left-hand rule.

E5D07 What determines the strength of the magnetic field around a conductor?

A. The resistance divided by the current
B. The ratio of the current to the resistance
C. The diameter of the conductor
D. The amount of current flowing through the conductor

D Magnetic field strength is proportional to the current and is stronger when the current is greater.

E5D08 What type of energy is stored in an electromagnetic or electrostatic field?

A. Electromechanical energy
B. Potential energy
C. Thermodynamic energy
D. Kinetic energy

B Don't be confused — stored energy is always potential energy as long as it is stored.

E5D09 What happens to reactive power in an AC circuit that has both ideal inductors and ideal capacitors?

A. It is dissipated as heat in the circuit
B. It is repeatedly exchanged between the associated magnetic and electric fields, but is not dissipated
C. It is dissipated as kinetic energy in the circuit
D. It is dissipated in the formation of inductive and capacitive fields

B The reactive power moves back and forth between the magnetic and electric fields but is only stored and not dissipated. Only the resistive part of a circuit will dissipate power.

E5D10 How can the true power be determined in an AC circuit where the voltage and current are out of phase?

A. By multiplying the apparent power times the power factor
B. By dividing the reactive power by the power factor
C. By dividing the apparent power by the power factor
D. By multiplying the reactive power times the power factor

A Power factor (PF) is a quantity that relates the apparent power in a circuit to the real power. You can find the real or true power by multiplying the apparent power by the power factor.

E5D11 What is the power factor of an R-L circuit having a 60 degree phase angle between the voltage and the current?

A. 1.414
B. 0.866
C. 0.5
D. 1.73

C The power factor is also the cosine of the phase angle between the voltage and the current, and the cosine of 60° is 0.5. The cosine of 0° is 1. When the voltage and current are in phase the power factor is 1 and there is no reactive power — it is all real power. When the voltage and current are out of phase by 90°, the power is all reactive and the power factor is 0 (the cosine of 90°).

E5D12 How many watts are consumed in a circuit having a power factor of 0.2 if the input is 100 VAC at 4 amperes?

    A. 400 watts
    B. 80 watts
    C. 2000 watts
    D. 50 watts

B   Apply the power factor to compute the power from

$$P = E \times I \times PF = 100 \text{ V} \times 4 \text{ A} \times 0.2 = 80 \text{ W}$$

E5D13 How much power is consumed in a circuit consisting of a 100 ohm resistor in series with a 100 ohm inductive reactance drawing 1 ampere?

    A. 70.7 watts
    B. 100 watts
    C. 141.4 watts
    D. 200 watts

B   Start by finding the impedance and phase angle

$$|Z| = \sqrt{100^2 + 100^2} = 141 \, \Omega$$

and

$$\theta = \tan^{-1}\left(\frac{100}{100}\right) = 45°$$

Power factor (PF) = $\cos \theta = \cos 45° = 0.707$

Apparent power = $I^2Z = 1^2 \times 141 = 141$ watts

To find real power, multiply by PF

$$P = 141 \times 0.707 = 100 \text{ watts}$$

E5D14 What is reactive power?

    A. Wattless, nonproductive power
    B. Power consumed in wire resistance in an inductor
    C. Power lost because of capacitor leakage
    D. Power consumed in circuit Q

A   Reactive power is out-of-phase, nonproductive power created by inductors and capacitors that cause voltage and current to be out of phase by some amount.

E5D15 What is the power factor of an R-L circuit having a 45 degree phase angle between the voltage and the current?

    A. 0.866
    B. 1.0
    C. 0.5
    D. 0.707

D Use the formula

Power factor (PF) = cos 45° = 0.707

E5D16 What is the power factor of an R-L circuit having a 30 degree phase angle between the voltage and the current?

    A. 1.73
    B. 0.5
    C. 0.866
    D. 0.577

C Use the formula

Power factor (PF) = cos 30° = 0.866

E5D17 How many watts are consumed in a circuit having a power factor of 0.6 if the input is 200 VAC at 5 amperes?

    A. 200 watts
    B. 1000 watts
    C. 1600 watts
    D. 600 watts

D Use the formula

Real power = E × I × PF = 200 × 5 × 0.6 = 600 watts

E5D18 How many watts are consumed in a circuit having a power factor of 0.71 if the apparent power is 500 VA?

    A. 704 W
    B. 355 W
    C. 252 W
    D. 1.42 mW

B Use the formula

Real power = Apparent Power × PF = 500 × 0.71 = 355 watts

Circuit Components

There will be six questions on your Extra class examination from the Circuit Components subelement. These six questions will be taken from the six groups of questions labeled E6A through E6F.

E6A **Semiconductor materials and devices: semiconductor materials; germanium, silicon, P-type, N-type; transistor types: NPN, PNP, junction, field-effect transistors: enhancement mode; depletion mode; MOS; CMOS; N-channel; P-channel**

E6A01 In what application is gallium arsenide used as a semiconductor material in preference to germanium or silicon?

A. In high-current rectifier circuits
B. In high-power audio circuits
C. At microwave frequencies
D. At very low frequency RF circuits

C Gallium arsenide (GaAs) has performance advantages for use at microwave frequencies. For that reason, it is often used to make solid-state devices for operation on those frequencies.

E6A02 Which of the following semiconductor materials contains excess free electrons?

A. N-type
B. P-type
C. Bipolar
D. Insulated gate

A Since electrons carry a negative charge, symbolized by N, a semiconductor material that contains more free electrons than pure materials is called N-type material.

E6A03 Why does a PN-junction diode not conduct current when reverse biased?

  A. Only P-type semiconductor material can conduct current
  B. Only N-type semiconductor material can conduct current
  C. Holes in P-type material and electrons in the N-type material are separated by the applied voltage, widening the depletion region
  D. Excess holes in P-type material combine with the electrons in N-type material converting the entire diode into an insulator

C   Reverse bias moves the charge carriers (electrons and holes) away from the PN junction, widening the depletion region and preventing current flow.

E6A04   What is the name given to an impurity atom that adds holes to a semiconductor crystal structure?

  A. Insulator impurity
  B. N-type impurity
  C. Acceptor impurity
  D. Donor impurity

C   A hole represents a missing electron in the crystal structure. Holes easily accept free electrons to replace the missing electron. For that reason an impurity atom that adds holes to a semiconductor crystal structure is called an acceptor impurity.

E6A05   What is the alpha of a bipolar junction transistor?

  A. The change of collector current with respect to base current
  B. The change of base current with respect to collector current
  C. The change of collector current with respect to emitter current
  D. The change of collector current with respect to gate current

C   Alpha (α) is the ratio of collector current to emitter current. The smaller the base current, the closer the collector current comes to being equal to that of the emitter and the closer alpha comes to being 1.

E6A06   What is the beta of a bipolar junction transistor?

  A. The frequency at which the current gain is reduced to 1
  B. The change in collector current with respect to base current
  C. The breakdown voltage of the base to collector junction
  D. The switching speed of the transistor

B   Beta (β) is the ratio of collector current to base current. It is a measure of the transistor's current gain.

E6A07   Which of the following indicates that a silicon NPN junction transistor is biased on?

    A. Base-to-emitter resistance of approximately 6 to 7 ohms
    B. Base-to-emitter resistance of approximately 0.6 to 0.7 ohms
    C. Base-to-emitter voltage of approximately 6 to 7 volts
    D. Base-to-emitter voltage of approximately 0.6 to 0.7 volts

D   When an NPN transistor is biased on, base-to-emitter voltage (V_{BE}) should measure 0.6 to 0.75 V from base to emitter with the positive voltmeter lead connected to the base.

E6A08   What term indicates the frequency at which the grounded-base current gain of a transistor has decreased to 0.7 of the gain obtainable at 1 kHz?

    A. Corner frequency
    B. Alpha rejection frequency
    C. Beta cutoff frequency
    D. Alpha cutoff frequency

D   The alpha cutoff frequency of a transistor is the frequency at which a transistor's current gain in the grounded-base configuration has decreased to 0.7 of the gain obtainable at 1 kHz.

E6A09   What is a depletion-mode FET?

    A. An FET that exhibits a current flow between source and drain when no gate voltage is applied
    B. An FET that has no current flow between source and drain when no gate voltage is applied
    C. Any FET without a channel
    D. Any FET for which holes are the majority carriers

A   A depletion-mode FET has a channel that passes current from the source to the drain when there is no bias voltage applied between the gate and source. In operation, the gate of a depletion-mode FET is reversed biased (negative gate-to-source voltage). When the reverse bias is applied to the gate, the channel is depleted of charge carriers, and current decreases.

E6A10   In Figure E6-2, what is the schematic symbol for an N-channel dual-gate MOSFET?

    A. 2
    B. 4
    C. 5
    D. 6

B   Symbol 4 is a dual-gate N-channel MOSFET. The arrow points into the N-channel.

E6A11   In Figure E6-2, what is the schematic symbol for a P-channel junction FET?

    A. 1
    B. 2
    C. 3
    D. 6

A   In the figure, Symbol 1 is a P-channel junction FET. The arrow points out of the P-type channel.

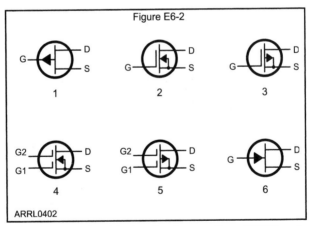

Figure E6-2 — Use this drawing for questions E6A10 and E6A11.

E6A12 Why do many MOSFET devices have internally connected Zener diodes on the gates?

    A. To provide a voltage reference for the correct amount of reverse-bias gate voltage
    B. To protect the substrate from excessive voltages
    C. To keep the gate voltage within specifications and prevent the device from overheating
    D. To reduce the chance of the gate insulation being punctured by static discharges or excessive voltages

D Nearly all the MOSFETs manufactured today have built-in gate-protective Zener diodes. Without this protection, the gate insulation can be perforated easily by small static charges on your hand or by the application of excessive voltages to the device.

E6A13 What do the initials CMOS stand for?

    A. Common Mode Oscillating System
    B. Complementary Mica-Oxide Silicon
    C. Complementary Metal-Oxide Semiconductor
    D. Common Mode Organic Silicon

C Sometimes both P- and N-channel MOSFETs are placed on the same wafer. The resulting transistor arrays can be interconnected on the wafer and are designed to perform a variety of special functions. This construction is called complementary metal-oxide semiconductor (CMOS) because the P- and N-channel transistors operate in complementary ways.

E6A14 How does DC input impedance at the gate of a field-effect transistor compare with the DC input impedance of a bipolar transistor?

    A. They are both low impedance
    B. An FET has low input impedance; a bipolar transistor has high input impedance
    C. An FET has high input impedance; a bipolar transistor has low input impedance
    D. They are both high impedance

C The modes of operation are different for the two types of transistor and so are their input impedances. The FET has a high input impedance. By contrast, the bipolar transistor has a low input impedance.

E6A15 Which semiconductor material contains excess holes in the outer shell of electrons?

    A. N-type
    B. P-type
    C. Superconductor-type
    D. Bipolar-type

B   A hole represents a missing electron in the crystal structure. Holes easily accept free electrons to replace the missing electron and so act as the majority carriers of current. For that reason an impurity atom that adds holes to a semiconductor crystal structure is called an acceptor impurity.

E6A16 What are the majority charge carriers in N-type semiconductor material?

    A. Holes
    B. Free electrons
    C. Free protons
    D. Free neutrons

B   Since N-type semiconductor material contains more free electrons than pure materials, the majority carriers of current in N-type material are the free electrons.

E6A17 What are the names of the three terminals of a field-effect transistor?

    A. Gate 1, gate 2, drain
    B. Emitter, base, collector
    C. Emitter, base 1, base 2
    D. Gate, drain, source

D   The gate of an FET is the control terminal or element. The source and drain are the two ends of the conducting channel. Current flow through the channel is controlled by voltage applied between the gate and source.

E6B   Diodes

E6B01   What is the most useful characteristic of a Zener diode?
A. A constant current drop under conditions of varying voltage
B. A constant voltage drop under conditions of varying current
C. A negative resistance region
D. An internal capacitance that varies with the applied voltage

B   Zener diodes operate with current flowing in the reverse direction from regular diodes. They are specially manufactured to exhibit a constant voltage drop over a wide range of currents. A Zener diode can be used as a voltage regulator because of this ability to maintain a constant voltage.

E6B02   What is an important characteristic of a Schottky diode as compared to an ordinary silicon diode when used as a power supply rectifier?
A. Much higher reverse voltage breakdown
B. Controlled reverse avalanche voltage
C. Enhanced carrier retention time
D. Less forward voltage drop

D   The Schottky barrier is a special metal-semiconductor junction. A diode made with a Schottky barrier as opposed to the usual P-N junction has lower forward voltage drop and can rectify high frequency signals more effectively.

E6B03   What special type of diode is capable of both amplification and oscillation?
A. Point contact
B. Zener
C. Tunnel
D. Junction

C   Because of the negative resistance characteristics of tunnel diodes, they can be used to make both amplifiers and oscillators. Tunnel diodes are no longer used in modern electronics.

E6B04 What type of semiconductor device is designed for use as a voltage-controlled capacitor?

A. Varactor diode
B. Tunnel diode
C. Silicon-controlled rectifier
D. Zener diode

A Junction diodes exhibit an appreciable internal capacitance due to the regions of charge separated by the narrow depletion layer. It is possible to change the internal capacitance of a diode by varying the amount of reverse bias applied across the junction. Manufacturers have designed varactor diodes to take advantage of this property.

E6B05 What characteristic of a PIN diode makes it useful as an RF switch or attenuator?

A. Extremely high reverse breakdown voltage
B. Ability to dissipate large amounts of power
C. Reverse bias controls its forward voltage drop
D. A large region of intrinsic material

D PIN stands for P-type – Intrinsic (pure or undoped) – N-type material and describes the construction of a PIN diode. DC bias current through the diode controls the resistivity of the thick layer of intrinsic material to the RF current flow. This makes the diode useful as a variable resistor or switch at RF.

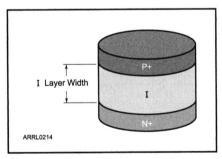

Figure E6B05 — This diagram illustrates the inner structure of a PIN diode. The top and bottom layers are labeled P+ and N+ to indicate very heavy levels of doping impurities are used.

E6B06 **Which of the following is a common use of a hot-carrier diode?**

A. As balanced mixers in FM generation
B. As a variable capacitance in an automatic frequency control circuit
C. As a constant voltage reference in a power supply
D. As a VHF / UHF mixer or detector

D Hot-carrier diodes have low internal capacitance and good high-frequency characteristics. They are often used in mixers and detectors at VHF and UHF.

E6B07 **What is the failure mechanism when a junction diode fails due to excessive current?**

A. Excessive inverse voltage
B. Excessive junction temperature
C. Insufficient forward voltage
D. Charge carrier depletion

B Junction temperature limits the maximum forward current rating in a junction diode. As current flow through a junction diode increases, the junction temperature will rise. If too much current flows, the junction gets too hot and the diode is destroyed.

E6B08 **Which of the following describes a type of semiconductor diode?**

A. Metal-semiconductor junction
B. Electrolytic rectifier
C. CMOS-field effect
D. Thermionic emission diode

A A metal-semiconductor junction is called a Schottky barrier. This type of diode is used as a high-speed rectifier.

E6B09 **What is a common use for point contact diodes?**

A. As a constant current source
B. As a constant voltage source
C. As an RF detector
D. As a high voltage rectifier

C Point-contact diodes have much less internal capacitance than PN-junction diodes. This means that point-contact diodes are better suited for RF applications. They are frequently used in RF detection circuits.

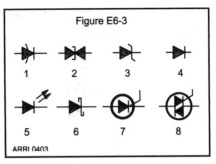

Figure E6-3

1 2 3 4

5 6 7 8

ARRL 0403

Figure E6-3 — Use this drawing for question E6B10.

E6B10 In Figure E6-3, what is the schematic symbol for a light-emitting diode?

A. 1
B. 5
C. 6
D. 7

B The schematic symbol for a light-emitting diode is shown in the drawing at number 5.

E6B11 What is used to control the attenuation of RF signals by a PIN diode?

A. Forward DC bias current
B. A sub-harmonic pump signal
C. Reverse voltage larger than the RF signal
D. Capacitance of an RF coupling capacitor

A See E6B05. DC bias current through the diode controls the resistivity of the thick layer of intrinsic material to the RF current flow. This makes the diode useful as a variable resistor or switch at RF.

E6B12 What is one common use for PIN diodes?

A. As a constant current source
B. As a constant voltage source
C. As an RF switch
D. As a high voltage rectifier

C See E6B05. PIN stands for P-type – Intrinsic (pure or undoped) – N-type material and describes the construction of a PIN diode. DC bias current through the diode controls the resistivity of the thick layer of intrinsic material to the RF current flow. This makes the diode particularly useful as a switch at RF. If you start building or repairing radios, at some point you will probably encounter a PIN diode being used as an RF switch.

E6B13 What type of bias is required for an LED to emit light?

A. Reverse bias
B. Forward bias
C. Zero bias
D. Inductive bias

B The LED emits light when the junction is forward biased and current flows through it.

E6C Digital ICs: families of digital ICs; gates; programmable logic devices (PLDs)

E6C01 What is the function of hysteresis in a comparator?

A. To prevent input noise from causing unstable output signals
B. To allow the comparator to be used with AC input signals
C. To cause the output to change states continually
D. To increase the sensitivity

A See E6C02. Hysteresis shifts the comparator's threshold by a small amount when the output changes state. This keeps noisy signals or signals close to the threshold from causing the output to switch repeatedly.

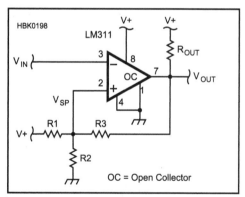

Figure E6C01 — A comparator circuit in which the output voltage is low when voltage at the inverting input is higher than the setpoint voltage, V_{SP}, at the noninverting input. R3 creates hysteresis by allowing more current to flow through R1 when the comparator output is low, shifting the setpoint by a few millivolts.

E6C02 What happens when the level of a comparator's input signal crosses the threshold?

A. The IC input can be damaged
B. The comparator changes its output state
C. The comparator enters latch-up
D. The feedback loop becomes unstable

B The function of a comparator is to indicate which of its input signals is greater. The threshold is the point at which both signals are equal. One input signal is often fixed and used as a reference or setpoint. When the remaining input signal crosses the threshold established by the setpoint, the comparator output state changes.

E6C03 What is tri-state logic?

A. Logic devices with 0, 1, and high impedance output states
B. Logic devices that utilize ternary math
C. Low power logic devices designed to operate at 3 volts
D. Proprietary logic devices manufactured by Tri-State Devices

A The ability to place an output in a high impedance state, effectively disconnecting it, allows tri-state logic devices to share a common data or memory bus under the control of a microprocessor or other device that controls access to a shared resource.

E6C04 What is the primary advantage of tri-state logic?

A. Low power consumption
B. Ability to connect many device outputs to a common bus
C. High speed operation
D. More efficient arithmetic operations

B The ability to place an output in a high impedance state, effectively disconnecting it, allows tri-state logic devices to share a common data or memory bus under the control of a microprocessor or other device that controls access to a shared resource.

E6C05 What is an advantage of CMOS logic devices over TTL devices?

A. Differential output capability
B. Lower distortion
C. Immune to damage from static discharge
D. Lower power consumption

D CMOS logic devices are generally smaller, consume less power, and have a lower cost than other logic families. CMOS devices do not have a differential output.

E6C06 Why do CMOS digital integrated circuits have high immunity to noise on the input signal or power supply?

A. Larger bypass capacitors are used in CMOS circuit design
B. The input switching threshold is about two times the power supply voltage
C. The input switching threshold is about one-half the power supply voltage
D. Input signals are stronger

C CMOS logic levels typically are within 0.1 V of the supply levels. In other words, if you use a 9 V supply the high level will be in the range of 8.9 to 9 V. In the same circumstances a low level will be from 0 to 0.1 V. The input switching threshold is about one-half the power supply voltage, which in this case is 4.5 V. In this example, noise would have to be greater than about 4.4 V to upset circuit operation. This is a substantially higher margin for noise than other logic families, particularly those operating with lower supply voltages.

E6C07 What best describes a pull up or pull down resistor?

A. A resistor in a keying circuit used to reduce key clicks
B. A resistor connected to the positive or negative supply line used to establish a voltage when an input or output is an open circuit
C. A resistor that insures that an oscillator frequency does not drive lower over time
D. A resistor connected to an op-amp output that only functions when the logic output is false

B Connecting a digital IC input or output to a fixed voltage, such as a power supply or circuit common, through a resistor forces an open-circuited input or output to a known logic level.

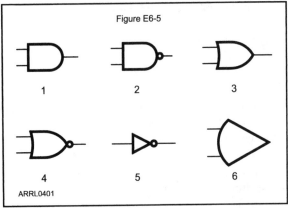

Figure E6-5 — Use this diagram for questions E6C08, E6C10, and E6C11.

E6C08 In Figure E6-5, what is the schematic symbol for a NAND gate?

A. 1
B. 2
C. 3
D. 4

B The schematic symbol for a NAND gate is symbol number 2 in the drawing.

E6C09 What is a Programmable Logic Device (PLD)?

A. A device to control industrial equipment
B. A programmable collection of logic gates and circuits in a single integrated circuit
C. Programmable equipment used for testing digital logic integrated circuits
D. An algorithm for simulating logic functions during circuit design

B By programming the connections between the various gates and circuits, complex and customized circuits can be created in the PLD.

E6C10 In Figure E6-5, what is the schematic symbol for a NOR gate?

A. 1
B. 2
C. 3
D. 4

D The schematic symbol for a NOR gate is shown in the drawing at 4.

E6C11   In Figure E6-5, what is the schematic symbol for the NOT operation (inverter)?

    A. 2
    B. 4
    C. 5
    D. 6

C   The schematic symbol for the NOT operation (inverter) is shown at 5.

E6C12   What is BiCMOS logic?

    A. A logic device with two CMOS circuits per package
    B. A FET logic family based on bimetallic semiconductors
    C. A logic family based on bismuth CMOS devices
    D. An integrated circuit logic family using both bipolar and CMOS transistors

D   Manufacturing processes that create both bipolar and CMOS devices on a single substrate allow the construction of integrated circuits that combine the best features of both types of transistors.

E6C13   Which of the following is an advantage of BiCMOS logic?

    A. Its simplicity results in much less expensive devices than standard CMOS
    B. It is totally immune to electrostatic damage
    C. It has the high input impedance of CMOS and the low output impedance of bipolar transistors
    D. All of these choices are correct

C   See E6C12.

E6C14   What is the primary advantage of using a Programmable Gate Array (PGA) in a logic circuit?

    A. Many similar gates are less expensive than a mixture of gate types
    B. Complex logic functions can be created in a single integrated circuit
    C. A PGA contains its own internal power supply
    D. All of these choice are correct

B   See E6C09.

E6D Toroidal and solenoidal inductors: permeability, core material, selecting, winding; transformers; piezoelectric devices

E6D01 How many turns will be required to produce a 5-microhenry inductor using a powdered-iron toroidal core that has an inductance index (A L) value of 40 microhenries/100 turns?

    A. 35 turns
    B. 13 turns
    C. 79 turns
    D. 141 turns

A Note that the units given for A_L in the question (inductance/turn) should be inductance/turns-squared.

$$N = 100 \sqrt{\frac{L}{A_L}}$$

where L = the inductance in mH; A_L = the inductance index, in mH per 100 turns²; and N = the number of turns.

Plug the values into the equation and you get

$$N = 100 \sqrt{\frac{L}{A_L}} = 100 \sqrt{\frac{5}{40}} = 100 \sqrt{0.125} = 35 \text{ turns}$$

E6D02 What is the equivalent circuit of a quartz crystal?

    A. Motional capacitance, motional inductance and loss resistance in series, all in parallel with a shunt capacitor representing electrode and stray capacitance
    B. Motional capacitance, motional inductance, loss resistance, and a capacitor representing electrode and stray capacitance all in parallel
    C. Motional capacitance, motional inductance, loss resistance, and a capacitor representing electrode and stray capacitance all in series
    D. Motional inductance and loss resistance in series, paralleled with motional capacitance and a capacitor representing electrode and stray capacitance

A Electrically, a quartz crystal acts as a parallel RLC circuit with a very high Q so that it oscillates at a single frequency with a very high stability. When placed in a holder, the resulting equivalent circuit is shown in Figure E6D02. L, C, and R are the electrical equivalents of the crystal's mechanical properties and CH is the capacitance of the holder plates with the crystal acting as the dielectric.

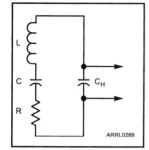

Figure E6D02 — The electrical equivalent circuit of a quartz crystal in a holder. L, C and R are the electrical equivalents of the crystal's mechanical properties and CH is the capacitance of the holder plates, with the crystal serving as the dielectric.

E6D03   Which of the following is an aspect of the piezoelectric effect?

    A. Mechanical deformation of material by the application of a voltage

    B. Mechanical deformation of material by the application of a magnetic field

    C. Generation of electrical energy in the presence of light

    D. Increased conductivity in the presence of light

A   Piezoelectric materials convert mechanical stress to a voltage and vice versa. This results from deformation of the material's crystalline lattice structure.

E6D04   Which materials are commonly used as a slug core in a variable inductor?

    A. Polystyrene and polyethylene

    B. Ferrite and brass

    C. Teflon and Delrin

    D. Cobalt and aluminum

B   Ferrite and brass cores are used for their abilities to concentrate or reduce (respectively) the magnetic field in an inductor.

E6D05   What is one reason for using ferrite cores rather than powdered-iron in an inductor?

    A. Ferrite toroids generally have lower initial permeability

    B. Ferrite toroids generally have better temperature stability

    C. Ferrite toroids generally require fewer turns to produce a given inductance value

    D. Ferrite toroids are easier to use with surface mount technology

C   Ferrite material generally has a much higher permeability than that of powdered iron. For that reason, ferrite toroids generally require fewer turns to produce a given inductance value.

E6D06 What core material property determines the inductance of a toroidal inductor?

    A. Thermal impedance
    B. Resistance
    C. Reactivity
    D. Permeability

D   The inductance of any inductor is determined by the number of turns of wire around its core and by the permeability of the core material. Permeability refers to the strength of a magnetic field in the core as compared to the strength of the field if no core were used. Cores with higher values of permeability will produce larger inductance values for the same number of turns on the core.

E6D07 What is the usable frequency range of inductors that use toroidal cores, assuming a correct selection of core material for the frequency being used?

    A. From a few kHz to no more than 30 MHz
    B. From less than 20 Hz to approximately 300 MHz
    C. From approximately 10 Hz to no more than 3000 kHz
    D. From about 100 kHz to at least 1000 GHz

B   By careful selection of core material, it is possible to produce toroidal inductors that can be used over the range of 20 Hz to around 300 MHz.

E6D08 What is one reason for using powdered-iron cores rather than ferrite cores in an inductor?

    A. Powdered-iron cores generally have greater initial permeability
    B. Powdered-iron cores generally maintain their characteristics at higher currents
    C. Powdered-iron cores generally require fewer turns to produce a given inductance
    D. Powdered-iron cores use smaller diameter wire for the same inductance

B   The choice of core materials for a particular inductor requires a compromise of characteristics. Ferrite cores generally have higher permeability values. Powdered-iron cores generally can be operated at higher currents. Since current flow can cause the core to heat as current flows, temperature stability is improved in powdered-iron cores.

E6D09 What devices are commonly used as VHF and UHF parasitic suppressors at the input and output terminals of a transistor HF amplifier?

A. Electrolytic capacitors
B. Butterworth filters
C. Ferrite beads
D. Steel-core toroids

C A ferrite bead is a very small core with a hole designed to slip over a component lead. These are often used as parasitic suppressors at the input and output terminals of transistorized HF amplifiers.

E6D10 What is a primary advantage of using a toroidal core instead of a solenoidal core in an inductor?

A. Toroidal cores confine most of the magnetic field within the core material
B. Toroidal cores make it easier to couple the magnetic energy into other components
C. Toroidal cores exhibit greater hysteresis
D. Toroidal cores have lower Q characteristics

A A primary advantage of using a toroidal core to wind an inductor rather than a linear core is that nearly all the magnetic field is contained within the core of the toroid. With a linear core, the magnetic field extends into the space surrounding the inductor. A toroidal core inductor can be placed near conducting surfaces and other coils with little effect on its performance.

E6D11 How many turns will be required to produce a 1-mH inductor using a ferrite toroidal core that has an inductance index (A L) value of 523 millihenries/1000 turns?

A. 2 turns
B. 4 turns
C. 43 turns
D. 229 turns

C Note that the units given for A_L in the question (inductance/turn) should be inductance/turns-squared. The equation to use for ferrite cores is

$$N = 1000 \sqrt{\frac{L}{A_L}}$$

where L = the inductance in mH; A_L = the inductance index, in mH per 1000 turns^2; and N = the number of turns.

Plugging the values into the equation gives

$$N = 1000 \sqrt{\frac{L}{A_L}} = 1000 \sqrt{\frac{1}{523}} = 1000 \sqrt{1.91 \times 10^{-3}} = 43.7 \text{ turns}$$

Which is approximately 43 turns.

E6D12   What is the definition of saturation in a ferrite core inductor?

A. The inductor windings are over coupled
B. The inductor's voltage rating is exceeded causing a flashover
C. The ability of the inductor's core to store magnetic energy is exceeded
D. Adjacent inductors become over-coupled

C   When the core becomes saturated, it no longer responds as an inductor to increases in current. This is true for any core material.

E6D13   What is the primary cause of inductor self-resonance?

A. Inter-turn capacitance
B. The skin effect
C. Inductive kickback
D. Non-linear core hysteresis

A   The capacitance between the various surfaces of the inductor act like a capacitance in series with the inductor, creating a series resonance called the self-resonant frequency.

E6D14   Which type of slug material decreases inductance when inserted into a coil?

A. Ceramic
B. Brass
C. Ferrite
D. Powdered-iron

B   Because the permeability of brass is less than that of air (relative permeability of 0.442), it reduces inductance when inserted into an inductor. This makes brass cores useful in adjustable inductors.

E6D15   What is current in the primary winding of a transformer called if no load is attached to the secondary?

A. Magnetizing current
B. Direct current
C. Excitation current
D. Stabilizing current

A   Magnetizing current establishes the magnetic field in the transformer core independently of additional current required to supply power to the secondary circuit.

E6D16 What is the common name for a capacitor connected across a transformer secondary that is used to absorb transient voltage spikes?

A. Clipper capacitor
B. Trimmer capacitor
C. Feedback capacitor
D. Snubber capacitor

D Snubber or suppressor capacitors store the energy of transients (short pulses) to protect other circuits.

E6D17 Why should core saturation of a conventional impedance matching transformer be avoided?

A. Harmonics and distortion could result
B. Magnetic flux would increase with frequency
C. RF susceptance would increase
D. Temporary changes of the core permeability could result

A The nonlinearities in current and voltage caused by saturation distort the applied waveform which creates harmonics and other distortion products.

E6E   Analog ICs: MMICs, CCDs, device packages

E6E01 Which of the following is true of a charge-coupled device (CCD)?

A. Its phase shift changes rapidly with frequency
B. It is a CMOS analog-to-digital converter
C. It samples an analog signal and passes it in stages from the input to the output
D. It is used in a battery charger circuit

C The analog signal fills the CCD input cell (called a charge bucket) with charge which is then passed or coupled to subsequent cells. A microprocessor measures the amount of charge in each cell to analyze the input signal.

E6E02 Which of the following device packages is a through-hole type?

A. DIP
B. PLCC
C. Ball grid array
D. SOT

A The Dual In-line Package (DIP) consists of two rows of pins that are inserted through holes in a printed circuit board for soldering to the board.

E6E03 Which of the following materials is likely to provide the highest frequency of operation when used in MMICs?

A. Silicon
B. Silicon nitride
C. Silicon dioxide
D. Gallium nitride

D The key to high operating frequency for semiconductors is the speed at which electrons can move through the material. Several advanced compounds, one of which is gallium nitride (GaN), have very high electron mobility and so are used for ICs and other semiconductor components that operate at microwave frequencies.

E6E04 What is the most common input and output impedance of circuits that use MMICs?

A. 50 ohms
B. 300 ohms
C. 450 ohms
D. 10 ohms

A Since MMICs frequently interface with other RF devices and systems, they are designed with a characteristic impedance of 50 ohms for the input and the output.

E6E05 Which of the following noise figure values is typical of a low-noise UHF preamplifier?

A. 2 dB
B. –10 dB
C. 44 dBm
D. –20 dBm

A Noise figure measures the amount of noise generated internally by a circuit or receiver so lower noise figures mean less noise will be added to the signal. Lower noise figures generally increase system sensitivity, particularly at VHF and higher frequencies. Noise figure values are expressed in dB.

E6E06 What characteristics of the MMIC make it a popular choice for VHF through microwave circuits?

A. The ability to retrieve information from a single signal even in the presence of other strong signals.
B. Plate current that is controlled by a control grid
C. Nearly infinite gain, very high input impedance, and very low output impedance
D. Controlled gain, low noise figure, and constant input and output impedance over the specified frequency range

D Because all of these attributes are combined in a small and inexpensive package, MMICs are very easy to use. MMICs can be used to replace many discrete components and reduce equipment size and weight, as well.

E6E07 Which of the following is typically used to construct a MMIC-based microwave amplifier?

A. Ground-plane construction
B. Microstrip construction
C. Point-to-point construction
D. Wave-soldering construction

B MMICs are surface mount devices and MMIC construction typically uses microstrip techniques. Double-sided circuit board material is used, and one side serves to form a ground plane for the circuit. Circuit traces form sections of feed line. The line widths, along with the circuit-board thickness and dielectric constant of the insulating material determine the characteristic impedance, which is normally 50 ohms.

E6E08 How is voltage from a power supply normally furnished to the most common type of monolithic microwave integrated circuit (MMIC)?

A. Through a resistor and/or RF choke connected to the amplifier output lead
B. MMICs require no operating bias
C. Through a capacitor and RF choke connected to the amplifier input lead
D. Directly to the bias-voltage (VCC IN) lead

A The operating bias voltage is normally supplied to a four-lead MMIC through a decoupling resistor or RF choke connected to the amplifier output lead. This technique is illustrated in Figure E6E08 which shows how MMICs are connected in a practical circuit. Power is supplied to the output pin through decoupling circuits and dc blocking capacitors allow the output signal from one MMIC to be connected to the input of the next.

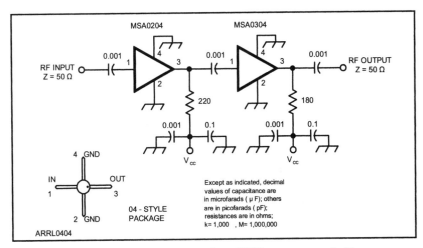

Figure E6E08 — This circuit illustrates how MMICs are connected in a practical circuit. Power is supplied to the output pin through decoupling circuits. DC blocking capacitors all the output signal from one MMIC to be connected to the input of the next MMIC.

E6E09 Which of the following component package types would be most suitable for use at frequencies above the HF range?

A. TO-220
B. Axial lead
C. Radial lead
D. Surface-mount

D Surface-mount packages have no wire leads, largely avoiding the effects of lead inductance. (See also E5D02.)

E6E10 What is the packaging technique in which leadless components are soldered directly to circuit boards?

A. Direct soldering
B. Virtual lead mounting
C. Stripped lead
D. Surface mount

D Surface-mount packages are placed on the printed-circuit board and soldered directly to pads without leads.

E6E11 What is a characteristic of DIP packaging used for integrated circuits?

A. Package mounts in a direct inverted position
B. Low leakage doubly insulated package
C. Two chips in each package (Dual In Package)
D. A total of two rows of connecting pins placed on opposite sides of the package (Dual In-line Package)

D DIP package pins are laid out in rows to assist with printed-circuit board layout and automated part placement.

E6E12 Why are high-power RF amplifier ICs and transistors sometimes mounted in ceramic packages?

A. High-voltage insulating ability
B. Better dissipation of heat
C. Enhanced sensitivity to light
D. To provide a low-pass frequency response

B Packaging for high-power devices must conduct heat well in order to pass it to a heat sink. At the same time, the package should be an electrical insulator — a tough combination. Some ceramics have the necessary properties so they are found in the packages of high-power RF transistors and ICs.

E6F   Optical components: photoconductive principles and effects, photovoltaic systems, optical couplers, optical sensors, and optoisolators; LCDs

E6F01   What is photoconductivity?

A. The conversion of photon energy to electromotive energy
B. The increased conductivity of an illuminated semiconductor
C. The conversion of electromotive energy to photon energy
D. The decreased conductivity of an illuminated semiconductor

B   The total conductance of a material may increase and its resistance decrease when light shines on the surface. This is called the photoconductive effect, or photoconductivity.

E6F02   What happens to the conductivity of a photoconductive material when light shines on it?

A. It increases
B. It decreases
C. It stays the same
D. It becomes unstable

A   The conductance of a photoconductive material increases when light shines on it.

E6F03   What is the most common configuration of an optoisolator or optocoupler?

A. A lens and a photomultiplier
B. A frequency modulated helium-neon laser
C. An amplitude modulated helium-neon laser
D. An LED and a phototransistor

D   An optoisolator or optocoupler is an LED and a phototransistor in a single IC package. Light from the LED shines on the phototransistor, allowing current to flow in the output circuit.

E6F04   What is the photovoltaic effect?

A. The conversion of voltage to current when exposed to light
B. The conversion of light to electrical energy
C. The conversion of electrical energy to mechanical energy
D. The tendency of a battery to discharge when used outside

B   The photovoltaic effect describes the creation of voltage or electric current in a material when exposed to light. When photons of light are absorbed by electrons in the material, they move to the outer shells of atoms where they can conduct electricity and this creates a voltage across the material.

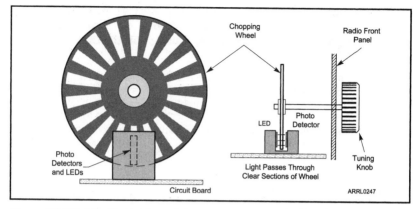

Figure E6F05 – An illustration of the operation of an optical shaft encoder, often used for the tuning and control functions of transceivers.

E6F05   Which describes an optical shaft encoder?

A. A device which detects rotation of a control by interrupting a light source with a patterned wheel
B. A device which measures the strength a beam of light using analog to digital conversion
C. A digital encryption device often used to encrypt spacecraft control signals
D. A device for generating RTTY signals by means of a rotating light source.

A   An optical shaft encoder usually consists of two pairs of emitters and detectors. A plastic disc with alternating clear and black radial bands rotates through a gap between the emitters and detectors. By using two emitters and two detectors, a microprocessor can detect the rotation direction and speed of the wheel. Modern transceivers use optical encoders to control the frequency of a synthesized VFO, for example.

E6F06   Which of these materials is affected the most by photoconductivity?

A. A crystalline semiconductor
B. An ordinary metal
C. A heavy metal
D. A liquid semiconductor

A   Crystalline semiconductors work best. Metals can be affected by photoconductivity but not as much as semiconductors. Liquid semiconductors are affected less than crystalline semiconductors.

E6F07   What is a solid state relay?

A. A relay using transistors to drive the relay coil
B. A device that uses semiconductors to implement the functions of an electromechanical relay
C. A mechanical relay that latches in the on or off state each time it is pulsed
D. A passive delay line

B   Solid-state relays (SSRs) use FETs or bipolar transistors to switch ac or dc currents under the control of a digital signal. SSRs do not use an electromagnet to move physical contacts so they can operate with much lower power and much higher speed than an electromagnetic relay.

E6F08   Why are optoisolators often used in conjunction with solid state circuits when switching 120 VAC?

A. Optoisolators provide a low impedance link between a control circuit and a power circuit
B. Optoisolators provide impedance matching between the control circuit and power circuit
C. Optoisolators provide a very high degree of electrical isolation between a control circuit and the circuit being switched
D. Optoisolators eliminate the effects of reflected light in the control circuit

C   They are called optoisolators because they have a very high impedance between the input and the output. This provides a high degree of electrical isolation between the controlling circuit and the controlled circuit.

E6F09   What is the efficiency of a photovoltaic cell?

A. The output RF power divided by the input dc power
B. The effective payback period
C. The open-circuit voltage divided by the short-circuit current under full illumination
D. The relative fraction of light that is converted to current

D   Efficiency is an important characteristic of photovoltaic cells or panels because it determines how much power can be obtained from a fixed amount of light.

E6F10   What is the most common type of photovoltaic cell used for electrical power generation?

A. Selenium
B. Silicon
C. Cadmium sulfide
D. Copper oxide

B   Most photovoltaic or solar cells are made from crystalline silicon, the same material used in most transistors. Even though it does not have the highest efficiency, it is inexpensive because of its wide use.

E6F11   Which of the following is the approximate open-circuit voltage produced by a fully-illuminated silicon photovoltaic cell?

A. 0.1 V
B. 0.5 V
C. 1.5 V
D. 12 V

B   Open-circuit means the output voltage with no load connected. In full sunlight, a typical silicon solar cell will develop 0.5 V at its output terminals.

E6F12   What absorbs the energy from light falling on a photovoltaic cell?

A. Protons
B. Photons
C. Electrons
D. Holes

C   When an electron near a semiconductor solar cell's depletion region absorbs a photon, the additional energy enables it to travel across the junction and combine with a hole to generate electricity.

E6F13   What is a liquid-crystal display (LCD)?

A. A modern replacement for a quartz crystal oscillator which displays its fundamental frequency
B. A display using a crystalline liquid which, in conjunction with polarizing filters, becomes opaque when voltage is applied
C. A frequency-determining unit for a transmitter or receiver
D. A display that uses a glowing liquid to remain brightly lit in dim light

B   LCDs place a thin layer of liquid crystals between a glass panel with transparent thin electrodes printed on it and a backplane and light emitting or reflecting material. When voltage is applied between the backplane and electrodes, the liquid crystal molecules twist to become opaque, creating a pattern.

E6F14   Which of the following is true of LCD displays?

A. They are hard to view in high ambient light conditions
B. They may be hard to view through polarized lenses
C. They only display alphanumeric symbols
D. All of these choices are correct

B   See E6F13. Because the liquid crystals polarize light themselves when twisted, that can render the pattern invisible if viewed through polarizing (Polaroid) material with the opposite polarization.

Practical Circuits

There will be eight questions on your Extra class examination from the Practical Circuits subelement. These eight questions will be taken from the eight groups of questions labeled E7A through E7H.

E7A Digital circuits: digital circuit principles and logic circuits: classes of logic elements; positive and negative logic; frequency dividers; truth tables

E7A01 Which is a bistable circuit?

A. An AND gate
B. An OR gate
C. A flip-flop
D. A clock

C Bistable means that the circuit has two (bi) stable states, such as on or off. Bistable multivibrator is just another name for a digital logic flip-flop. For that reason, a flip-flop can be used to store one bit of information.

E7A02 What is the function of a decade counter digital IC?

A. It produces one output pulse for every ten input pulses
B. It decodes a decimal number for display on a seven-segment LED display
C. It produces ten output pulses for every input pulse
D. It adds two decimal numbers together

A A decade counter produces one output pulse for every ten input pulses. Counter circuits are also called dividers.

E7A03   Which of the following can divide the frequency of a pulse train by 2?

A. An XOR gate
B. A flip-flop
C. An OR gate
D. A multiplexer

B   See E7A01. A flip-flop, also known as a bistable multivibrator, can be used as a frequency divider. Since the output state changes once for each input pulse, it takes two pulses to complete the output cycle from one binary state, to the opposite state, then back to the original state.

E7A04   How many flip-flops are required to divide a signal frequency by 4?

A. 1
B. 2
C. 4
D. 8

B   Each flip-flop has the ability to divide a signal's frequency by 2, so the answer to this question is 4/2 = 2.

E7A05   Which of the following is a circuit that continuously alternates between two states without an external clock?

A. Monostable multivibrator
B. J-K flip-flop
C. T flip-flop
D. Astable multivibrator

D   The prefix "a" means "not" so an astable (not stable) multivibrator alternates between two unstable states. Ordinarily an unstable circuit is undesirable but the astable multivibrator is useful as a digital oscillator to generate clock or other timing signals.

E7A06   What is a characteristic of a monostable multivibrator?

A. It switches momentarily to the opposite binary state and then returns to its original state after a set time
B. It is a clock that produces a continuous square wave oscillating between 1 and 0
C. It stores one bit of data in either a 0 or 1 state
D. It maintains a constant output voltage, regardless of variations in the input voltage

A   A monostable multivibrator, meaning "one stable state" and also called a "single-shot" or "one-shot," switches momentarily to the opposite binary state and then returns after a set time to its original state. The length of the momentary period is determined by circuit component values.

Table E7-1

Output of Various Logic Gates

Input 1	Input 2	AND	NAND	OR	NOR	XOR
0	0	0	1	0	1	1
0	1	0	1	1	0	0
1	0	0	1	1	0	0
1	1	1	0	1	0	1

E7A07 What logical operation does a NAND gate perform?

A. It produces a logic "0" at its output only when all inputs are logic "0"
B. It produces a logic "1" at its output only when all inputs are logic "1"
C. It produces a logic "0" at its output if some but not all of its inputs are logic "1"
D. It produces a logic "0" at its output only when all inputs are logic "1"

D Table E7-1 shows that a NAND (NOT-AND) gate produces a logic 0 at its output only when all inputs are logic 1.

E7A08 What logical operation does an OR gate perform?

A. It produces a logic "1" at its output if any or all inputs are logic "1"
B. It produces a logic "0" at its output if all inputs are logic "1"
C. It only produces a logic "0" at its output when all inputs are logic "1"
D. It produces a logic "1" at its output if all inputs are logic "0"

A Table E7-1 shows that an OR gate produces a logic 1 at its output if either or both of its inputs are logic 1.

E7A09 What logical operation is performed by an exclusive NOR gate?

A. It produces a logic "0" at its output only if all inputs are logic "0"
B. It produces a logic "1" at its output only if all inputs are logic "1"
C. It produces a logic "0" at its output if any single input is a logic "1"
D. It produces a logic "1" at its output if any single input is a logic "1"

C Table E7-1 shows that an exclusive NOR gate (XNOR) produces a logic 1 at its output if just one of its inputs are logic 1.

E7A10 What is a truth table?

A. A table of logic symbols that indicate the high logic states of an op-amp
B. A diagram showing logic states when the digital device output is true
C. A list of inputs and corresponding outputs for a digital device
D. A table of logic symbols that indicates the logic states of an op-amp

C Table E7-1 is an example of a truth table. A truth table lists input combinations and their corresponding outputs for a digital device or devices. In other words, the truth table shows all of the combinations that are true for that device.

E7A11 What type of logic defines "1" as a high voltage?

A. Reverse Logic
B. Assertive Logic
C. Negative logic
D. Positive Logic

D In positive logic, a 1 is represented by a high voltage level.

E7A12 What type of logic defines "0" as a high voltage?

A. Reverse Logic
B. Assertive Logic
C. Negative logic
D. Positive Logic

C In a negative-logic circuit, a low level is used to represent a logic 1.

E7B Amplifiers: class of operation; vacuum tube and solid-state circuits; distortion and intermodulation; spurious and parasitic suppression; microwave amplifiers; switching-type amplifiers

E7B01 For what portion of a signal cycle does a Class AB amplifier operate?

A. More than 180 degrees but less than 360 degrees
B. Exactly 180 degrees
C. The entire cycle
D. Less than 180 degrees

A The Class AB amplifier conducts for more than 180 degrees (Class B) but less than 360 degrees (Class A).

E7B02 What is a Class D amplifier?

A. A type of amplifier that uses switching technology to achieve high efficiency
B. A low power amplifier using a differential amplifier for improved linearity
C. An amplifier using drift-mode FETs for high efficiency
D. A frequency doubling amplifier

A The Class D amplifier operates similarly to a switching power supply in that the amplifying transistor is operated as a switch that supplies energy to an output filter circuit. The switch is opened and closed at a very high rate with respect to the signal being amplified so that the output filter can remove the undesired signal components that result from the pulses of energy.

E7B03 Which of the following components form the output of a class D amplifier circuit?

A. A low-pass filter to remove switching signal components
B. A high-pass filter to compensate for low gain at low frequencies
C. A matched load resistor to prevent damage by switching transients
D. A temperature compensating load resistor to improve linearity

A To remove the undesired signal components that result from the pulses of energy from the amplifying device, the output filter must be a low-pass filter.

E7B04   Where on the load line of a Class A common emitter amplifier would bias normally be set?

A. Approximately half-way between saturation and cutoff
B. Where the load line intersects the voltage axis
C. At a point where the bias resistor equals the load resistor
D. At a point where the load line intersects the zero bias current curve

A   A load line describes the possible combinations of output voltage and current. One end of the line is at zero current and full voltage (cutoff). The other lies at full current and zero voltage (saturation). The amplifier circuit's design determines where the operating point falls along that line. Class A power amplifiers operate between the saturation or cutoff points for linear amplification so that the amplifying device is conducting current at all times.

E7B05   What can be done to prevent unwanted oscillations in an RF power amplifier?

A. Tune the stage for maximum SWR
B. Tune both the input and output for maximum power
C. Install parasitic suppressors and/or neutralize the stage
D. Use a phase inverter in the output filter

C   Neutralization is a technique of using negative feedback to prevent parasitic oscillations in a power amplifier caused by positive feedback within the amplifying devices.

E7B06   Which of the following amplifier types reduces or eliminates even-order harmonics?

A. Push-push
B. Push-pull
C. Class C
D. Class AB

B   The push-pull configuration with two amplifying devices operated in Class B reduces even-order harmonics.

E7B07   Which of the following is a likely result when a Class C amplifier is used to amplify a single-sideband phone signal?

A. Reduced intermodulation products
B. Increased overall intelligibility
C. Signal inversion
D. Signal distortion and excessive bandwidth

D   Class C amplifiers are highly nonlinear and nonlinear amplifiers cause distortion. The distorted waveform contains harmonics and spurious signals (splatter) that occupy a wide bandwidth. SSB requires linear amplification to be received properly and to occupy the minimum bandwidth necessary.

E7B08   How can an RF power amplifier be neutralized?

A. By increasing the driving power
B. By reducing the driving power
C. By feeding a 180-degree out-of-phase portion of the output back to the input
D. By feeding an in-phase component of the output back to the input

C   Neutralization is a technique of using negative feedback to prevent parasitic oscillations in a power amplifier caused by positive feedback within the amplifying devices. Negative feedback is out-of-phase with the input signal and acts to stabilize the amplifier gain.

E7B09   Which of the following describes how the loading and tuning capacitors are to be adjusted when tuning a vacuum tube RF power amplifier that employs a Pi-network output circuit?

A. The loading capacitor is set to maximum capacitance and the tuning capacitor is adjusted for minimum allowable plate current
B. The tuning capacitor is set to maximum capacitance and the loading capacitor is adjusted for minimum plate permissible current
C. The loading capacitor is adjusted to minimum plate current while alternately adjusting the tuning capacitor for maximum allowable plate current
D. The tuning capacitor is adjusted for minimum plate current, while the loading capacitor is adjusted for maximum permissible plate current

D   The procedure for tuning a vacuum-tube power amplifier having an output pi-network is to alternately increase the plate current with the loading capacitor and dip the plate current with the tuning capacitor. The goal is to produce the required amount of output power with the minimum amount of plate current.

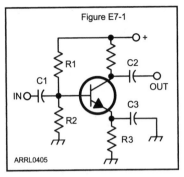

Figure E7-1 — Use this diagram for question E7B10, E7B11, and E7B12.

E7B10   In Figure E7-1, what is the purpose of R1 and R2?

A. Load resistors
B. Fixed bias
C. Self bias
D. Feedback

B   The two resistors form a voltage divider that provides a fixed bias voltage and current for the base.

E7B11   In Figure E7-1, what is the purpose of R3?

A. Fixed bias
B. Emitter bypass
C. Output load resistor
D. Self bias

D   Current through R3 creates a voltage drop across the resistor, which acts to oppose the bias provided by R1 and R2. This stabilizes the transistor's collector current and is called self-bias.

E7B12   What type of amplifier circuit is shown in Figure E7-1?

A. Common base
B. Common collector
C. Common emitter
D. Emitter follower

C   Since the emitter is bypassed to ground and the output is taken from the collector circuit, it is called a common-emitter amplifier.

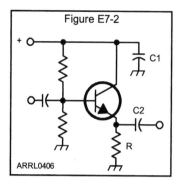

Figure E7-2

Figure E7-2 — Use this diagram for question E7B13.

ARRL0406

E7B13   In Figure E7-2, what is the purpose of R?

    A. Emitter load
    B. Fixed bias
    C. Collector load
    D. Voltage regulation

A   Since the resistor is connected to the emitter and the collector is bypassed to ground, it represents the emitter load.

E7B14   Why are switching amplifiers more efficient than linear amplifiers?

    A. Switching amplifiers operate at higher voltages
    B. The power transistor is at saturation or cut off most of the time, resulting in low power dissipation
    C. Linear amplifiers have high gain resulting in higher harmonic content
    D. Switching amplifiers use push-pull circuits

B   When the switching transistor is fully on (saturated) or off (cutoff), it dissipates little power as a switch. By keeping the transistor in this state as much as possible, the amplifier can be more efficient than a linear amplifier.

E7B15   What is one way to prevent thermal runaway in a bipolar transistor amplifier?

    A. Neutralization
    B. Select transistors with high beta
    C. Use a resistor in series with the emitter
    D. All of these choices are correct

C   Degenerative emitter feedback is another term for the self-bias created by a emitter resistor. The feedback consists of the voltage drop across the emitter. It is degenerative because as the voltage increases from higher emitter current, it acts to reduce the amount of bias and collector current. Degenerative feedback is a type of negative feedback.

E7B16 What is the effect of intermodulation products in a linear power amplifier?

    A. Transmission of spurious signals
    B. Creation of parasitic oscillations
    C. Low efficiency
    D. All of these choices are correct

A   The mixing products that result from intermodulation create additional sidebands on frequencies near the desired signal. These spurious signals interfere with communications on these frequencies.

E7B17 Why are odd-order rather than even-order intermodulation distortion products of concern in linear power amplifiers?

    A. Because they are relatively close in frequency to the desired signal
    B. Because they are relatively far in frequency from the desired signal
    C. Because they invert the sidebands causing distortion
    D. Because they maintain the sidebands, thus causing multiple duplicate signals

A   Third-order intermodulation products from nonlinear amplification of complex signals such as SSB phone signals are near the original frequency. For example, a two-tone SSB signal with components f_1 and f_2 generates third-order intermodulation products at $2f_1 \pm f_2$ and $2f_2 \pm f_1$. These are close enough to the original signals to interfere with communications in the same band.

E7B18 What is a characteristic of a grounded-grid amplifier?

    A. High power gain
    B. High filament voltage
    C. Low input impedance
    D. Low bandwidth

C   Although other types of amplifier circuits have higher power gain, grounded-grid amplifiers are popular because their low input impedance is easy to match to the 50 ohms of transmitter output circuits. They also have a wide operating bandwidth of useful power gain.

E7C   Filters and matching networks: types of networks; types of filters; filter applications; filter characteristics; impedance matching; DSP filtering

E7C01   How are the capacitors and inductors of a low-pass filter Pi-network arranged between the network's input and output?

A. Two inductors are in series between the input and output, and a capacitor is connected between the two inductors and ground

B. Two capacitors are in series between the input and output, and an inductor is connected between the two capacitors and ground

C. An inductor is connected between the input and ground, another inductor is connected between the output and ground, and a capacitor is connected between the input and output

D. A capacitor is connected between the input and ground, another capacitor is connected between the output and ground, and an inductor is connected between input and output

D   The correct configuration should look like the Greek letter π (pi). Since the question is looking for the low pass filter, we want the one with the inductor in the middle. This configuration is low-pass because capacitors pass higher frequencies and block low frequencies while inductors pass low frequencies and block high frequencies. This shunts the high frequencies to ground and blocks them from passing to the output. In the high-pass filter pi-network, the low frequencies are shunted to ground while the high frequencies pass through.

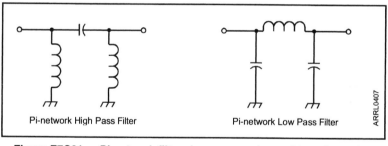

Pi-network High Pass Filter   Pi-network Low Pass Filter

Figure E7C01 — Pi-network filters have one series and two shunt elements.

E7C02 Which of the following is a property of a T-network with series capacitors and a parallel shunt inductor?

- A. It is a low-pass filter
- B. It is a band-pass filter
- C. It is a high-pass filter
- D. It is a notch filter

C The T-network transforms impedances as a pair of back-to-back L-networks. It is a high pass filter because the inductor shunts the low frequencies to ground while the capacitors allow the high frequencies to pass through. Figure E7C02 shows three types of impedance matching networks: (A) is an L-network, (B) is a Pi-network, and (C) is a T-network. A and B are shown in a low-pass configuration. The T-network at C acts as a high-pass filter.

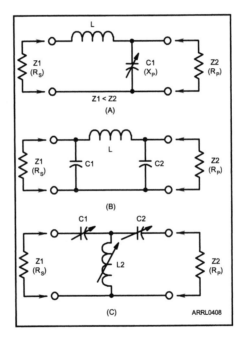

Figure E7C02 — Three types of impedance matching networks. (A) is an L-network, (B) is a Pi-network, and (C) is a T-network. A and B are shown in a low-pass configuration. The T-network at C acts as a high-pass filter.

E7C03 What advantage does a Pi-L-network have over a Pi-network for impedance matching between the final amplifier of a vacuum-tube transmitter and an antenna?

  A. Greater harmonic suppression
  B. Higher efficiency
  C. Lower losses
  D. Greater transformation range

A   The additional inductor of the Pi-L network increases high frequency rejection at the cost of additional losses, reduced efficiency, and a somewhat reduced range of impedances that can be transformed.

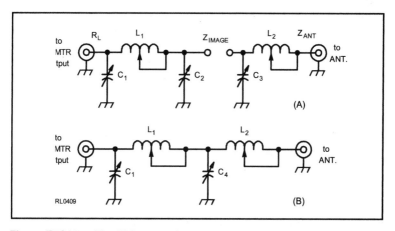

Figure E7C03 — The Pi-L network uses a Pi network to transform the transmitter output impedance (RL) to the image impedance (Z_{IMAGE}). An L-network then transforms the image impedance to the antenna impedance, Z_{ANT}. This is shown at A. Because C2 and C3 are in parallel, they are combined into a single capacitor (C4) as shown in the practical configuration at B.

E7C04   How does an impedance-matching circuit transform a complex impedance to a resistive impedance?

  A. It introduces negative resistance to cancel the resistive part of impedance
  B. It introduces transconductance to cancel the reactive part of impedance
  C. It cancels the reactive part of the impedance and changes the resistive part to a desired value
  D. Network resistances are substituted for load resistances and reactances are matched to the resistances

C   Impedance matching networks transform one impedance to another by changing the ratios of voltage and current (impedance) and by changing the phase relationship between the voltage and current (reactance).

E7C05 Which filter type is described as having ripple in the passband and a sharp cutoff?

A. A Butterworth filter
B. An active LC filter
C. A passive op-amp filter
D. A Chebyshev filter

D The Chebyshev filter design equations result in a filter with amplitude response variations (ripple) in the passband. This tradeoff is made to obtain a sharp cutoff. The figure shows typical filter response curves for Butterworth or maximally-flat (A), Chebyshev (B), and elliptical (C) designs.

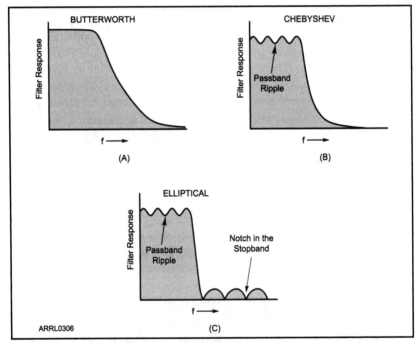

Figure E7C05 — Typical filter response curves for Butterworth or maximally-flat (A), Chebyshev (B), and elliptical (C) designs.

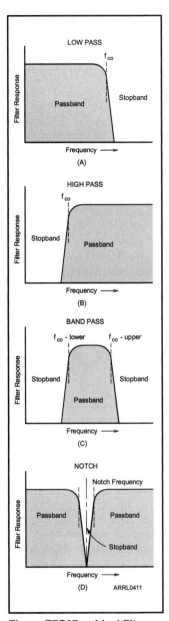

Figure E7C07 — Ideal filter response curves for low-pass, high-pass, band-pass, and notch filters.

E7C06 What are the distinguishing features of an elliptical filter?

A. Gradual passband rolloff with minimal stop band ripple
B. Extremely flat response over its pass band with gradually rounded stop band corners
C. Extremely sharp cutoff with one or more notches in the stop band
D. Gradual passband rolloff with extreme stop band ripple

C See E7C05. Like the Chebyshev filters, elliptical filters trade away the flatness of the amplitude response for sharper cutoff. The elliptical filter design also produces notches in the stop-band (the region in which the filter attenuates frequencies).

E7C07 What kind of filter would you use to attenuate an interfering carrier signal while receiving an SSB transmission?

A. A band-pass filter
B. A notch filter
C. A Pi-network filter
D. An all-pass filter

B A notch filter (or band-stop filter) attenuates a very narrow range of frequencies, just right for removing the single tone of interfering carriers while affecting the remainder of the signal as little as possible. The figure shows ideal filter response curves for low-pass, high-pass, band-pass, and notch filters.

E7C08 Which of the following factors has the greatest effect in helping determine the bandwidth and response shape of a crystal ladder filter?

A. The relative frequencies of the individual crystals
B. The DC voltage applied to the quartz crystal
C. The gain of the RF stage preceding the filter
D. The amplitude of the signals passing through the filter

A The bandwidth and response shape depend primarily on the resonant frequencies of the crystals used to construct the filter.

E7C09 What is a Jones filter as used as part of an HF receiver IF stage?

A. An automatic notch filter
B. A variable bandwidth crystal lattice filter
C. A special filter that emphasizes image responses
D. A filter that removes impulse noise

B The Jones filter is a special type of crystal filter whose bandwidth can be controlled. This provides an adjustable-bandwidth filter with the superior response of fixed-bandwidth crystal filters.

E7C10 Which of the following filters would be the best choice for use in a 2 meter repeater duplexer?

A. A crystal filter
B. A cavity filter
C. A DSP filter
D. An L-C filter

B Cavity filters are resonant enclosures with extremely low losses and thus high Q and narrow bandwidths. Cavity filters can be configured to pass or remove signals, making them especially useful for repeaters where the transmit and receive frequencies are quite close together.

E7C11 Which of the following is the common name for a filter network which is equivalent to two L networks connected back-to-back with the inductors in series and the capacitors in shunt at the input and output?

A. Pi-L
B. Cascode
C. Omega
D. Pi

D See E7C03. The Pi-network takes its name from its resemblance to the Greek letter π. A Pi-network consists of one inductor and two capacitors or two inductors and one capacitor. The T-network is another circuit that can be thought of as a pair of back-to-back L-networks.

E7C12 Which describes a Pi-L network used for matching a vacuum-tube final amplifier to a 50-ohm unbalanced output?

    A. A Phase Inverter Load network
    B. A Pi network with an additional series inductor on the output
    C. A network with only three discrete parts
    D. A matching network in which all components are isolated from ground

B   See E7C03. A Pi-L-network is essentially a Pi-network with an additional series element. The Pi-L network most common in amateur consists of two inductors as the series components and two capacitors as the shunt elements.

E7C13 What is one advantage of a Pi matching network over an L matching network consisting of a single inductor and a single capacitor?

    A. The Q of Pi-networks can be varied depending on the component values chosen
    B. L-networks cannot perform impedance transformation
    C. Pi-networks have fewer components
    D. Pi-networks are designed for balanced input and output

A   Pi-networks transform impedances in two steps, one by the input pair of components and one by the output pair of components. By choosing the transformation ratios of each stage, the Q of the overall circuit can also be adjusted.

E7C14 Which mode is most affected by non-linear phase response in a receiver IF filter?

    A. Meteor Scatter
    B. Single-Sideband voice
    C. Digital
    D. Video

C   Digital communications often use relative phase to encode the data symbols and so those modes are highly dependent on the phase response of filters.

E7C15 What is a crystal lattice filter?

    A. A power supply filter made with interlaced quartz crystals
    B. An audio filter made with four quartz crystals that resonate at 1-kHz intervals
    C. A filter with wide bandwidth and shallow skirts made using quartz crystals
    D. A filter with narrow bandwidth and steep skirts made using quartz crystals

D   Crystal lattice filters are made using quartz crystals. They are characterized as having a relatively narrow bandwidth and steep skirts.

E7D Power supplies and voltage regulators; solar array charge controllers

E7D01 What is one characteristic of a linear electronic voltage regulator?

    A. It has a ramp voltage as its output

    B. It eliminates the need for a pass transistor

    C. The control element duty cycle is proportional to the line or load conditions

    D. The conduction of a control element is varied to maintain a constant output voltage

D A linear voltage regulator varies the conductance of a control element (such as a transistor) to maintain a constant output voltage to the load. Because this is a linear circuit, the control is continuous. There are no switching elements used in this type of regulator.

E7D02 What is one characteristic of a switching electronic voltage regulator?

    A. The resistance of a control element is varied in direct proportion to the line voltage or load current

    B. It is generally less efficient than a linear regulator

    C. The control device's duty cycle is controlled to produce a constant average output voltage

    D. It gives a ramp voltage at its output

C A switching regulator, in contrast to a linear regulator has switching elements (such as transistors) that control the regulator output voltage. In switching regulators the control device is switched on and off electronically, with the duty cycle automatically adjusted to maintain a constant average output voltage. Switching frequencies of several kilohertz and higher are used to avoid the need for extensive filtering to remove ripple at the switching frequency from the dc output.

E7D03 What device is typically used as a stable reference voltage in a linear voltage regulator?

    A. A Zener diode
    B. A tunnel diode
    C. An SCR
    D. A varactor diode

A   A Zener diode can provide a stable voltage reference and is frequently used for that purpose in linear voltage regulators.

E7D04 Which of the following types of linear voltage regulator usually make the most efficient use of the primary power source?

    A. A series current source
    B. A series regulator
    C. A shunt regulator
    D. A shunt current source

B   The series regulator draws current from the primary power source in proportion to the load. In other words, it only draws power from the primary source when that power is needed.

E7D05 Which of the following types of linear voltage regulator places a constant load on the unregulated voltage source?

    A. A constant current source
    B. A series regulator
    C. A shunt current source
    D. A shunt regulator

D   The shunt regulator regulates output voltage by providing a constant load to the unregulated (primary) voltage source. When more power is needed in the load, less power is diverted through the shunt element (usually a transistor). When less power is needed in the load, more power is diverted to the shunt element of the regulator.

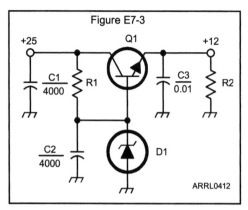

Figure E7-3 — Use this diagram for questions E7D06 through E7D08.

E7D06   What is the purpose of Q1 in the circuit shown in Figure E7-3?

A. It provides negative feedback to improve regulation
B. It provides a constant load for the voltage source
C. It increases the current-handling capability of the regulator
D. It provides D1 with current

C   The Zener diode will regulate voltage over a limited current range. Transistor Q1 increases that regulation range and thus increases the current-handling capability of the regulator circuit.

E7D07   What is the purpose of C2 in the circuit shown in Figure E7-3?

A. It bypasses hum around D1
B. It is a brute force filter for the output
C. To self-resonate at the hum frequency
D. To provide fixed DC bias for Q1

A   See E7D06. The capacitor is used to filter out any ac signal components from the reference voltage.

E7D08   What type of circuit is shown in Figure E7-3?

A. Switching voltage regulator
B. Grounded emitter amplifier
C. Linear voltage regulator
D. Emitter follower

C   See E7D06. The schematic in Figure E7-3 is of a linear voltage regulator.

E7D09 What is the main reason to use a charge controller with a solar power system?

A. Prevention of battery undercharge
B. Control of electrolyte levels during battery discharge
C. Prevention of battery damage due to overcharge
D. Matching of day and night charge rates

C Simply hooking a battery to a solar panel can ruin the battery through over-charge or damage the solar panel. A controller circuit prevents damage and maximizes battery life.

E7D10 What is the primary reason that a high-frequency switching type high-voltage power supply can be both less expensive and lighter in weight than a conventional power supply?

A. The inverter design does not require any output filtering
B. It uses a diode bridge rectifier for increased output
C. The high frequency inverter design uses much smaller transformers and filter components for an equivalent power output
D. It uses a large power-factor compensation capacitor to create free power from the unused portion of the AC cycle

C Because the cycles associated with the high-frequency inverter are much shorter than those of a 60-Hz linear supply, the transformers and capacitors can be much smaller and still perform equally well. The tradeoff is that the inverter circuit adds expense.

E7D11 What circuit element is controlled by a series analog voltage regulator to maintain a constant output voltage?

A. Reference voltage
B. Switching inductance
C. Error amplifier
D. Pass transistor

D The pass transistor acts as a variable resistor to allow just enough load current to maintain a constant voltage.

E7D12 What is the drop-out voltage of an analog voltage regulator?

A. Minimum input voltage for rated power dissipation
B. Maximum amount that the output voltage drops when the input voltage is varied over its specified range
C. Minimum input-to-output voltage required to maintain regulation
D. Maximum amount that the output voltage may decrease at rated load

C For the pass transistor to regulate output voltage a minimum amount of voltage drop is required from the unregulated supply to the regulated output. Below that voltage, the pass transistor can no longer maintain regulation and voltage falls or drops out.

E7D13 What is the equation for calculating power dissipation by a series connected linear voltage regulator?

A. Input voltage multiplied by input current
B. Input voltage divided by output current
C. Voltage difference from input to output multiplied by output current
D. Output voltage multiplied by output current

C The pass transistor dissipates power just like a resistor and the equation is the same: voltage across the transistor multiplied by the current through the transistor.

E7D14 What is one purpose of a "bleeder" resistor in a conventional unregulated power supply?

A. To cut down on waste heat generated by the power supply
B. To balance the low-voltage filament windings
C. To improve output voltage regulation
D. To boost the amount of output current

C Bleeder resistors provide a constant load to the supply, keeping the unregulated voltage within an acceptable range. Bleeder resistors also discharge the supply filter capacitors over a relatively long period. This is done for safety reasons.

E7D15 What is the purpose of a "step-start" circuit in a high-voltage power supply?

A. To provide a dual-voltage output for reduced power applications
B. To compensate for variations of the incoming line voltage
C. To allow for remote control of the power supply
D. To allow the filter capacitors to charge gradually

D Without a step-start or soft-start circuit in the supply, the discharged filter capacitors draw very high current at turn-on. This short-term current surge can be stressful to supply components, including the capacitors themselves. The step-start circuit reduces the surge by turning on the supply more gradually.

E7D16 When several electrolytic filter capacitors are connected in series to increase the operating voltage of a power supply filter circuit, why should resistors be connected across each capacitor?

    A. To equalize, as much as possible, the voltage drop across each capacitor

    B. To provide a safety bleeder to discharge the capacitors when the supply is off

    C. To provide a minimum load current to reduce voltage excursions at light loads

    D. All of these choices are correct

D   Capacitors with the same nominal value are not identical in actual value. The resulting charge imbalance between several connected in series leads to an unequal charge distribution and voltage. Using a resistors in parallel with the series-connected capacitors also equalizes the voltage across the capacitors, ensuring that the voltage across them remains within their ratings.

E7E   Modulation and demodulation: reactance, phase and balanced modulators; detectors; mixer stages

E7E01 Which of the following can be used to generate FM phone emissions?

    A. A balanced modulator on the audio amplifier

    B. A reactance modulator on the oscillator

    C. A reactance modulator on the final amplifier

    D. A balanced modulator on the oscillator

B   A reactance modulator connected to the transmitter master oscillator's tank circuit is a simple and satisfactory device for producing true FM in an amateur transmitter.

E7E02 What is the function of a reactance modulator?

    A. To produce PM signals by using an electrically variable resistance

    B. To produce AM signals by using an electrically variable inductance or capacitance

    C. To produce AM signals by using an electrically variable resistance

    D. To produce PM signals by using an electrically variable inductance or capacitance

D   A reactance modulator connected to the output of an oscillator circuit introduces a varying phase delay, creating a phase-modulated (PM) signal. A reactance modulator can also produce FM.

E7E03 How does an analog phase modulator function?

A. By varying the tuning of a microphone preamplifier to produce PM signals

B. By varying the tuning of an amplifier tank circuit to produce AM signals

C. By varying the tuning of an amplifier tank circuit to produce PM signals

D. By varying the tuning of a microphone preamplifier to produce AM signals

C A phase modulator varies the tuning of an RF amplifier tank circuit to produce PM signals by introducing a variable phase shift in the signal path.

E7E04 What is one way a single-sideband phone signal can be generated?

A. By using a balanced modulator followed by a filter

B. By using a reactance modulator followed by a mixer

C. By using a loop modulator followed by a mixer

D. By driving a product detector with a DSB signal

A You can generate a single-sideband phone signal by using a balanced modulator followed by a filter. The balanced modulator produces a double-sideband suppressed-carrier signal. The filter removes the unwanted sideband.

E7E05 What circuit is added to an FM transmitter to boost the higher audio frequencies?

A. A de-emphasis network

B. A heterodyne suppressor

C. An audio prescaler

D. A pre-emphasis network

D A pre-emphasis circuit is used in a transmitter to attenuate lower frequencies so that the modulation caused by high and low audio frequencies is equalized.

E7E06 Why is de-emphasis commonly used in FM communications receivers?

A. For compatibility with transmitters using phase modulation

B. To reduce impulse noise reception

C. For higher efficiency

D. To remove third-order distortion products

A Since phase modulation emphasizes high audio frequencies with a pre-emphasis (high-pass) filter a de-emphasis (low-pass) filter is required to restore the proper balance between high and low frequency components of the modulating signal.

E7E07   What is meant by the term baseband in radio communications?

A. The lowest frequency band that the transmitter or receiver covers
B. The frequency components present in the modulating signal
C. The unmodulated bandwidth of the transmitted signal
D. The basic oscillator frequency in an FM transmitter that is multiplied to increase the deviation and carrier frequency

B   Baseband signals can be analog signals such as voice or data signals.

E7E08   What are the principal frequencies that appear at the output of a mixer circuit?

A. Two and four times the original frequency
B. The sum, difference and square root of the input frequencies
C. The two input frequencies along with their sum and difference frequencies
D. 1.414 and 0.707 times the input frequency

C   In the mixing process, two signals are multiplied together. This produces mixing product signals at the sum and difference of the original two frequencies. In a practical mixer, output signals at the original two frequencies are also present, but at a reduced level.

E7E09   What occurs when an excessive amount of signal energy reaches a mixer circuit?

A. Spurious mixer products are generated
B. Mixer blanking occurs
C. Automatic limiting occurs
D. A beat frequency is generated

A   Spurious mixing products will be produced if input-signal energy overloads the mixer circuit. The level of these spurious mixing products may increase to the point that they are detectable in the output. One result of these effects is that the receiver may suffer severe interference in the presence of extremely strong signals.

E7E10   How does a diode detector function?

A. By rectification and filtering of RF signals
B. By breakdown of the Zener voltage
C. By mixing signals with noise in the transition region of the diode
D. By sensing the change of reactance in the diode with respect to frequency

A   A diode rectifies an AM signal, leaving the signal's envelope (containing the modulation information) and pulses at the AM carrier frequency. The envelope can be detected with a circuit that responds only to low frequencies, ignoring or filtering out the high-frequency pulses.

E7E11   Which type of detector is used for demodulating SSB signals?

  A. Discriminator
  B. Phase detector
  C. Product detector
  D. Phase comparator

C   A product detector mixes an incoming signal with a local signal or beat-frequency oscillator (BFO). This converts the SSB signal from RF to audio frequencies.

E7E12   What is a frequency discriminator stage in a FM receiver?

  A. An FM generator circuit
  B. A circuit for filtering two closely adjacent signals
  C. An automatic band-switching circuit
  D. A circuit for detecting FM signals

D   A frequency discriminator is found in an FM receiver where it functions as a detector, recovering the modulating information.

E7F   DSP filtering and other operations; software defined radio fundamentals; DSP modulation and demodulation

E7F01   What is meant by direct digital conversion as applied to software defined radios?

  A. Software is converted from source code to object code during operation of the receiver
  B. Incoming RF is converted to a control voltage for a voltage controlled oscillator
  C. Incoming RF is digitized by an analog-to-digital converter without being mixed with a local oscillator signal
  D. A switching mixer is used to generate I and Q signals directly from the RF input

C   Direct digital conversion samples the incoming RF and converts it to digital data without any frequency conversion as in a superheterodyne receiver.

E7F02   What kind of digital signal processing audio filter is used to remove unwanted noise from a received SSB signal?

  A. An adaptive filter
  B. A crystal-lattice filter
  C. A Hilbert-transform filter
  D. A phase-inverting filter

A   Because the DSP filter is created by software, it can sense the conditions at its input and adapt its operation accordingly. This allows a DSP adaptive filter to remove many types of interfering noise from voice and CW signals.

E7F03   What type of digital signal processing filter is used to generate an SSB signal?

A. An adaptive filter
B. A notch filter
C. A Hilbert-transform filter
D. An elliptical filter

C   The Hilbert-transform filter is difficult to implement with analog circuitry but straightforward with DSP. It creates a constant 90° phase shift independent of frequency and is applied to the baseband signal. Hilbert-transform filters are used in the phasing method of generating SSB which is the most common way of generating SSB with DSP.

E7F04   What is a common method of generating an SSB signal when using digital signal processing?

A. Mixing products are converted to voltages and subtracted by adder circuits
B. A frequency synthesizer removes the unwanted sidebands
C. Emulation of quartz crystal filter characteristics
D. Combine signals with a quadrature phase relationship

D   The phasing method which combines signals with a quadrature (90-degree) phase relationship is favorite for DSP-based equipment, whereas the phasing method is quite difficult for analog circuits.

E7F05   How frequently must an analog signal be sampled by an analog-to-digital converter so that the signal can be accurately reproduced?

A. At half the rate of the highest frequency component of the signal
B. At twice the rate of the highest frequency component of the signal
C. At the same rate as the highest frequency component of the signal
D. At four times the rate of the highest frequency component of the signal

B   Sampling a signal at twice the highest frequency to be reproduced is required according to the Nyquist Sampling Theorem.

E7F06   What is the minimum number of bits required for an analog-to-digital converter to sample a signal with a range of 1 volt at a resolution of 1 millivolt?

A. 4 bits
B. 6 bits
C. 8 bits
D. 10 bits

D   10 bits is 2 to the 10th (2^{10}) power or 1024 so the least significant bit (LSB) represents 1/1024th of the reference voltage or just less than 1 mV.

E7F07   What functions can a Fast Fourier Transform perform?

A. Converting analog signals to digital form
B. Converting digital signals to analog form
C. Converting digital signals from the time domain to the frequency domain
D. Converting 8-bit data to 16-bit data

C   The Fast Fourier Transform (FFT) is used to convert a series of samples representing a digitized analog signal (the time domain) to a series of values representing the spectrum of the analog signal at different frequencies (the frequency domain).

E7F08   What is the function of decimation with regard to digital filters?

A. Converting data to binary-coded decimal form
B. Reducing the effective sample rate by removing samples
C. Attenuating the signal
D. Removing unnecessary significant digits

B   Decimation consists of reducing a set of digital values to a fraction of the original size by removing every n-th sample or by removing all but every n-th sample, depending on how much sample set reduction is needed.

E7F09   Why is an anti-aliasing digital filter required in a digital decimator?

A. It removes high frequency signal components which would otherwise be reproduced as lower frequency components
B. It peaks the response of the decimator, improving bandwidth
C. It removes low frequency signal components to eliminate the need for DC restoration
D. It notches out the sampling frequency to avoid sampling errors

A   When the sample rate is reduced by decimation, signal components at frequencies higher than the reduced sample rate will cause alias. For this reason, they must be filtered out of the data before decimation is performed.

E7F10   What aspect of receiver analog-to-digital conversion determines the maximum receive bandwidth of a Direct Digital Conversion SDR?

A. Sample rate
B. Sample width in bits
C. Sample clock phase noise
D. Processor latency

A   See E7F05. The sample rate must be twice the received bandwidth or aliases will be produced, creating distortion and false signals.

E7F11 What sets the minimum detectable signal level for an SDR in the absence of atmospheric or thermal noise?

A. Sample clock phase noise
B. Reference voltage level and sample width in bits
C. Data storage transfer rate
D. Missing codes and jitter

B   See E7F06. The minimum detectable signal (MDS) for an SDR is determined by the size of the least significant bit (LSB). The size of the LSB is equal to the reference voltage divided by the 2 to the number of bits in each sample.

E7F12 What digital process is applied to I and Q signals to recover the baseband modulation information?

A. Fast Fourier Transform
B. Decimation
C. Signal conditioning
D. Quadrature mixing

A   Performing an Fast Fourier Transform (FFT) on digitized I and Q signals recovers the baseband spectrum of the modulating signal. An Inverse FFT and digital-to-analog conversion is then used to turn the information into the original analog signals.

E7F13 What is the function of taps in a digital signal processing filter?

A. To reduce excess signal pressure levels
B. Provide access for debugging software
C. Select the point at which baseband signals are generated
D. Provide incremental signal delays for filter algorithms

D   In a digital filter, each tap is a version of the input signal which is delayed by a fixed amount, different for each tap. Each tap output signal is then multiplied by a coefficient. The output of the taps are then added together to form the filtered signal.

E7F14 Which of the following would allow a digital signal processing filter to create a sharper filter response?

A. Higher data rate
B. More taps
C. Complex phasor representations
D. Double-precision math routines

B   The more taps a digital filter has, the more precise the filter's response can be. The tradeoff is that the signal delay through filter increases as more as taps are added.

E7F15 Which of the following is an advantage of a Finite Impulse Response (FIR) filter vs an Infinite Impulse Response (IIR) digital filter?

A. FIR filters delay all frequency components of the signal by the same amount
B. FIR filters are easier to implement for a given set of passband rolloff requirements
C. FIR filters can respond faster to impulses
D. All of these choices are correct

A The FIR filters are less-efficient than IIR filters but are guaranteed to have a stable output for any type of input signal.

E7F16 How might the sampling rate of an existing digital signal be adjusted by a factor of 3/4?

A. Change the gain by a factor of 3/4
B. Multiply each sample value by a factor of 3/4
C. Add 3 to each input value and subtract 4 from each output value
D. Interpolate by a factor of three, then decimate by a factor of four

D Multiplying the sample rate of a digital signal is done by interpolation while dividing the sample rate is done by decimation. Thus, to change the sample rate by 3/4, first multiply the rate by three through interpolation, then divide the sample rate by 4 through decimation.

E7F17 What do the letters I and Q in I/Q Modulation represent?

A. Inactive and Quiescent
B. Instantaneous and Quasi-stable
C. Instantaneous and Quenched
D. In-phase and Quadrature

D The I and Q signals have a 90° (quadrature) phase relationship. Modulated I and Q signals can be combined to create any kind of modulation to be created.

E7G   Active filters and op-amp circuits: active audio filters; characteristics; basic circuit design; operational amplifiers

E7G01 What is the typical output impedance of an integrated circuit op-amp?

A. Very low
B. Very high
C. 100 ohms
D. 1000 ohms

A The output impedance is the opposite of the input impedance — the ideal op amp has a very low output impedance.

E7G02   What is the effect of ringing in a filter?

   A. An echo caused by a long time delay
   B. A reduction in high frequency response
   C. Partial cancellation of the signal over a range of frequencies
   D. Undesired oscillations added to the desired signal

D   Filter ringing occurs when the Q of a filter is too high, creating narrow peaks in the filter's amplitude response. This creates oscillations in the filter when a signal is present at the frequency of the peak.

E7G03   What is the typical input impedance of an integrated circuit op-amp?

   A. 100 ohms
   B. 1000 ohms
   C. Very low
   D. Very high

D   The theoretical (perfect) op amp has infinite input impedance.

E7G04   What is meant by the term op-amp input offset voltage?

   A. The output voltage of the op-amp minus its input voltage
   B. The difference between the output voltage of the op-amp and the input voltage required in the immediately following stage
   C. The differential input voltage needed to bring the open-loop output voltage to zero
   D. The potential between the amplifier input terminals of the op-amp in an open-loop condition

C   Op amp input-offset voltage is the dc voltage between the input terminals required to bring the output to zero in an open-loop condition. It is measured by adding a differential voltage to the input terminals such that the output voltage is returned to zero. Offset results from imbalance between the IC's input transistors and their biasing circuits.

E7G05   How can unwanted ringing and audio instability be prevented in a multi-section op-amp RC audio filter circuit?

   A. Restrict both gain and Q
   B. Restrict gain, but increase Q
   C. Restrict Q, but increase gain
   D. Increase both gain and Q

A   To avoid unwanted ringing and audio instability in a multisection op amp RC audio filter circuit, you should limit (restrict) both the gain and the Q of the circuit.

E7G06   Which of the following is the most appropriate use of an op-amp active filter?

    A. As a high-pass filter used to block RFI at the input to receivers
    B. As a low-pass filter used between a transmitter and a transmission line
    C. For smoothing power-supply output
    D. As an audio filter in a receiver

D   Op amp RC active filters are well suited for audio-frequency applications.

E7G07   What magnitude of voltage gain can be expected from the circuit in Figure E7-4 when R1 is 10 ohms and RF is 470 ohms?

    A. 0.21
    B. 94
    C. 47
    D. 24

C   The voltage gain for the circuit in the drawing is

$$V_{GAIN} = \frac{-R_F}{R1} = \frac{-470}{10} = -47$$

The minus sign indicates that this is an inverting amplifier. That means that a positive input gives a negative output and vice versa. In stating gain, it is usual to ignore the inversion, and following that practice the gain is 47.

E7G08   How does the gain of an ideal operational amplifier vary with frequency?

    A. It increases linearly with increasing frequency
    B. It decreases linearly with increasing frequency
    C. It decreases logarithmically with increasing frequency
    D. It does not vary with frequency

D   In an ideal operational amplifier, there is no change in gain as the frequency changes. In other words, the frequency response is flat.

E7G09   What will be the output voltage of the circuit shown in Figure E7-4 if R1 is 1000 ohms, RF is 10,000 ohms, and 0.23 volts dc is applied to the input?

    A. 0.23 volts
    B. 2.3 volts
    C. -0.23 volts
    D. -2.3 volts

D   The output voltage is equal to voltage gain times the input voltage. So,

$$V_{OUT} = \frac{-R_F}{R1} \times V_{IN}$$

Using the values given,

$$V_{OUT} = \frac{-10,000}{1000} \times 0.23 = -2.3 \text{ V}$$

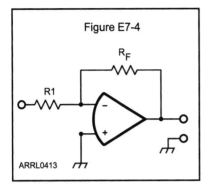

Figure E7-4

R_F

R1

ARRL0413

Figure E7-4 — Use this diagram to answer E7G07, E7G09, E7G10, and E7G11.

E7G10 What absolute voltage gain can be expected from the circuit in Figure E7-4 when R1 is 1800 ohms and RF is 68 kilohms?

    A. 1
    B. 0.03
    C. 38
    D. 76

C   Using the resistor values given

$$V_{GAIN} = \frac{-R_F}{R1} = \frac{-68,000}{1800} = -38$$

As before, you can ignore the minus sign when stating the circuit gain.

E7G11 What absolute voltage gain can be expected from the circuit in Figure E7-4 when R1 is 3300 ohms and RF is 47 kilohms?

    A. 28
    B. 14
    C. 7
    D. 0.07

B   Using the resistor values given, you'll find

$$V_{GAIN} = \frac{-R_F}{R1} = \frac{-47,000}{3300} = -14$$

Ignore the minus sign as you did before.

E7G12 **What is an integrated circuit operational amplifier?**

    A. A high-gain, direct-coupled differential amplifier with very high input impedance and very low output impedance

    B. A digital audio amplifier whose characteristics are determined by components external to the amplifier

    C. An amplifier used to increase the average output of frequency modulated amateur signals to the legal limit

    D. An RF amplifier used in the UHF and microwave regions

A   The operational amplifier (op amp) is a high-gain, direct-coupled, differential amplifier that will amplify dc signals as well as ac signals. The key item to spot here is differential amplifier.

E7H   **Oscillators and signal sources: types of oscillators; synthesizers and phase-locked loops; direct digital synthesizers; stabilizing thermal drift; microphonics; high accuracy oscillators**

E7H01   **What are three oscillator circuits used in Amateur Radio equipment?**

    A. Taft, Pierce and negative feedback

    B. Pierce, Fenner and Beane

    C. Taft, Hartley and Pierce

    D. Colpitts, Hartley and Pierce

D   Colpitts, Hartley and Pierce are the most popular oscillator circuits, named after their inventors. The figure shows the Hartley (A), Colpitts (B), and Pierce (C) circuits.

E7H02   **Which describes a microphonic?**

    A. An IC used for amplifying microphone signals

    B. Distortion caused by RF pickup on the microphone cable

    C. Changes in oscillator frequency due to mechanical vibration

    D. Excess loading of the microphone by an oscillator

C   If an oscillator responds to mechanical vibration by changing frequency, it is acting like a microphone. This response is called a microphonic.

E7H03   **How is positive feedback supplied in a Hartley oscillator?**

    A. Through a tapped coil

    B. Through a capacitive divider

    C. Through link coupling

    D. Through a neutralizing capacitor

A   See E7H01. Hartley starts with H, and so does henry — the basic unit of inductance. You can use that as a mnemonic to help you remember that the Hartley oscillator uses a tapped coil to provide the positive feedback needed for operation — circuit (A) in Figure E7H01.

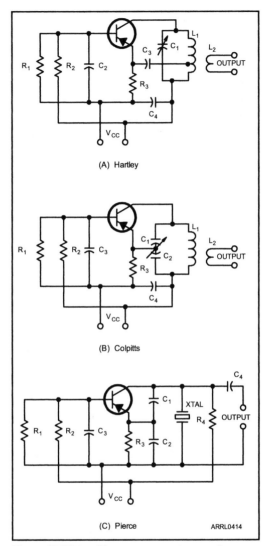

Figure E7H01 — Three common types of transistor oscillator circuits are the Hartley (A), Colpitts (B), and Pierce (C).

E7H04 How is positive feedback supplied in a Colpitts oscillator?

    A. Through a tapped coil
    B. Through link coupling
    C. Through a capacitive divider
    D. Through a neutralizing capacitor

C See E7H01. Colpitts starts with C, and so does capacitor. You can use that as a mnemonic to help you remember that the Colpitts oscillator uses a capacitive voltage divider to provide the positive feedback needed for operation — circuit (B) in Figure E7H01.

E7H05 How is positive feedback supplied in a Pierce oscillator?

    A. Through a tapped coil
    B. Through link coupling
    C. Through a neutralizing capacitor
    D. Through a quartz crystal

D See E7H01. The Pierce oscillator provides positive feedback through a quartz crystal — circuit (C) in Figure E7H01.

E7H06 Which of the following oscillator circuits are commonly used in VFOs?

    A. Pierce and Zener
    B. Colpitts and Hartley
    C. Armstrong and deForest
    D. Negative feedback and balanced feedback

B The right choice is Colpitts and Hartley but you could guess that by recognizing the other answers as all or partly not oscillators.

E7H07 How can an oscillator's microphonic responses be reduced?

    A. Use of NP0 capacitors
    B. Eliminating noise on the oscillator's power supply
    C. Using the oscillator only for CW and digital signals
    D. Mechanically isolating the oscillator circuitry from its enclosure

D See also E7H02. The best way to prevent microphonic responses is to prevent mechanical vibrations from reaching the oscillator.

E7H08 Which of the following components can be used to reduce thermal drift in crystal oscillators?

    A. NP0 capacitors
    B. Toroidal inductors
    C. Wirewound resistors
    D. Non-inductive resistors

A NP0 (Negative-Positive Zero) temperature coefficient capacitors are designed not to change value with temperature and so are used in crystal oscillators to reduce changes with temperature, called thermal drift.

E7H09 What type of frequency synthesizer circuit uses a phase accumulator, lookup table, digital to analog converter, and a low-pass anti-alias filter?

    A. A direct digital synthesizer
    B. A hybrid synthesizer
    C. A phase locked loop synthesizer
    D. A diode-switching matrix synthesizer

A   A direct digital synthesizer uses a phase accumulator, lookup table, digital to analog converter and a low-pass antialias filter. Figure E7H09 shows the block diagram of a direct digital synthesizer at (A). At (B), the amplitude values found in the ROM lookup table for a particular sine wave being generated by the synthesizer. The smoothed output signal from the synthesizer after it goes through the low-pass antialias filter is shown at (C).

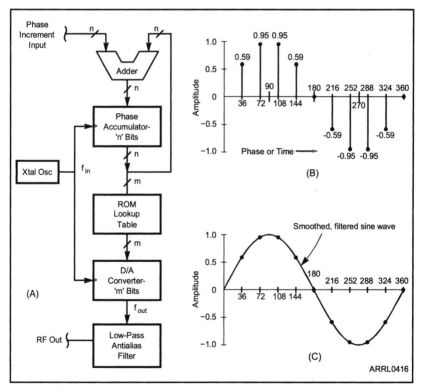

Figure E7H09 — The block diagram of a direct digital synthesizer is shown at (A). At (B), the amplitude values found in the ROM lookup table for a particular sine wave being generated by the synthesizer. The smoothed output signal from the synthesizer after it goes through the low-pass anti-aliasing filter is shown at (C).

E7H10 What information is contained in the lookup table of a direct digital frequency synthesizer?

    A. The phase relationship between a reference oscillator and the output waveform

    B. The amplitude values that represent a sine-wave output

    C. The phase relationship between a voltage-controlled oscillator and the output waveform

    D. The synthesizer frequency limits and frequency values stored in the radio memories

B   See E7H09. The lookup table of a direct digital frequency synthesizer is a list of sine-wave values at various phase angles.

E7H11 What are the major spectral impurity components of direct digital synthesizers?

    A. Broadband noise

    B. Digital conversion noise

    C. Spurious signals at discrete frequencies

    D. Nyquist limit noise

C   The major spectral impurity components are spurious signals, or spurs, at specific discrete frequencies related to the frequencies of the digital circuits that make up the synthesizer.

E7H12 Which of the following must be done to insure that a crystal oscillator provides the frequency specified by the crystal manufacturer?

    A. Provide the crystal with a specified parallel inductance

    B. Provide the crystal with a specified parallel capacitance

    C. Bias the crystal at a specified voltage

    D. Bias the crystal at a specified current

B   The capacitance of the crystal's holder or socket adds to the crystal's own capacitance so it must be accounted for when manufacturing the crystal to have a specific frequency.

E7H13 Which of the following is a technique for providing highly accurate and stable oscillators needed for microwave transmission and reception?

    A. Use a GPS signal reference

    B. Use a rubidium stabilized reference oscillator

    C. Use a temperature-controlled high Q dielectric resonator

    D. All of these choices are correct

D   The first two choices generate precise time and frequency signals from atomic clocks. The third choice keeps the resonator at a controlled temperature to avoid frequency changes from heating and cooling.

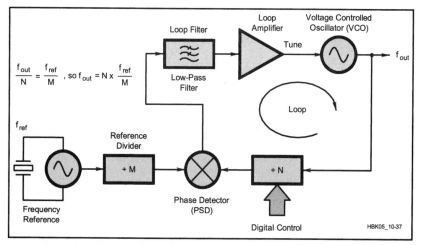

Figure E7H14 — A phase-locked loop (PLL) synthesizer acts to keep the divided-down signal from its voltage-controlled oscillator (VCO) phase-locked to the divided-down signal from its reference oscillator. By changing M and N, fine tuning steps in the VCO frequency can be made with the same frequency stability of the reference oscillator.

E7H14   What is a phase-locked loop circuit?

    A. An electronic servo loop consisting of a ratio detector, reactance modulator, and voltage-controlled oscillator

    B. An electronic circuit also known as a monostable multivibrator

    C. An electronic servo loop consisting of a phase detector, a low-pass filter, a voltage-controlled oscillator, and a stable reference oscillator

    D. An electronic circuit consisting of a precision push-pull amplifier with a differential input

C   A phase-locked loop (PLL) synthesizer uses a stable voltage-controlled oscillator, programmable divider, phase detector, loop filter and a reference frequency source.

E7H15   Which of these functions can be performed by a phase-locked loop?

    A. Wide-band AF and RF power amplification

    B. Comparison of two digital input signals, digital pulse counter

    C. Photovoltaic conversion, optical coupling

    D. Frequency synthesis, FM demodulation

D   See E7H14. Phase-locked loops (PLL) are used in frequency synthesis and FM demodulation applications. In the first example, variable frequency dividers are used so that the voltage-controlled oscillator operates at a selectable frequency. In the second, the loop's error signal contains the modulating signal as the PLL tracks the FM signal frequency.

Signals and Emissions

There will be four questions on your Extra class examination from the Signals and Emissions subelement. These four questions will be taken from the four groups of questions labeled E8A through E8D.

E8A   AC waveforms: sine, square, sawtooth and irregular waveforms; AC measurements; average and PEP of RF signals; Fourier analysis; analog to digital conversion: digital to analog conversion

E8A01   What is the name of the process that shows that a square wave is made up of a sine wave plus all of its odd harmonics?

A. Fourier analysis
B. Vector analysis
C. Numerical analysis
D. Differential analysis

A   By applying Fourier analysis, a waveform can be shown to be made up of multiple sine and cosine waveforms.

E8A02   What type of wave has a rise time significantly faster than its fall time (or vice versa)?

A. A cosine wave
B. A square wave
C. A sawtooth wave
D. A sine wave

C   Sawtooth waves are straight-line waveforms with unequal rise and fall times. Sawtooth waves consist of both odd and even harmonics, as well as the fundamental. A sawtooth wave appears on an oscilloscope as a straight-line waveform that has a rise time faster than the fall time (or vice-versa).

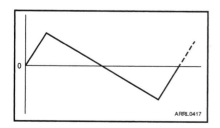

Figure E8A02 — Sawtooth waves are straight-line waveforms with unequal rise and fall times. Sawtooth waves consist of both odd and even harmonics, as well as the fundamental.

ARRL0417

E8A03 What type of wave does a Fourier analysis show to be made up of sine waves of a given fundamental frequency plus all of its harmonics?

A. A sawtooth wave
B. A square wave
C. A sine wave
D. A cosine wave

A   A sawtooth wave is made up of sine waves of a fundamental frequency and all harmonics. The figure shows how the waveform will look on a spectrum analyzer. You see components at 10 Hz, 20 Hz, 30 Hz, 40 Hz, 50 Hz, 60 Hz, 70 Hz, and on out to infinity. Since these are both the even and the odd multiples of the fundamental frequency (10 Hz), the sawtooth wave is said to be made up of the fundamental frequency and all harmonics.

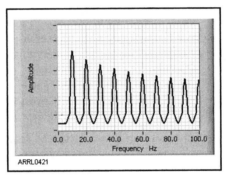

ARRL0421

Figure E8A03 — Spectrum analyzer display of a 10-Hz sawtooth wave. The sawtooth waveform's spectrum consists of components that include the fundamental at 10 Hz plus all of the harmonics at 20 Hz, 30 Hz, 40 Hz, and so on.

E8A04 What is "dither" with respect to analog to digital converters?

A. An abnormal condition where the converter cannot settle on a value to represent the signal
B. A small amount of noise added to the input signal to allow more precise representation of a signal over time
C. An error caused by irregular quantization step size
D. A method of decimation by randomly skipping samples

B   By adding a small amount of noise, measurements can avoid small offsets in the long-term average of the signal.

E8A05 What would be the most accurate way of measuring the RMS voltage of a complex waveform?

A. By using a grid dip meter
B. By measuring the voltage with a D'Arsonval meter
C. By using an absorption wavemeter
D. By measuring the heating effect in a known resistor

D When an ac voltage is applied to a resistor, the resistor will dissipate energy in the form of heat, just as if the voltage were dc. The dc voltage that would cause identical heating in the ac-excited resistor is called the root-mean-square (RMS) or effective value of the ac voltage.

E8A06 What is the approximate ratio of PEP-to-average power in a typical single-sideband phone signal?

A. 2.5 to 1
B. 25 to 1
C. 1 to 1
D. 100 to 1

A PEP (peak envelope power) is the average power during one cycle at a modulation peak. These envelope peaks occur sporadically during voice transmission. In Figure E8A06, you can see a couple of examples of PEP to average power in SSB signals. In (B), the average level is greater, which raises the average output power compared to the peak value. The ratio of peak-to-average amplitude varies widely with voices of different characteristics. The PEP of an SSB signal may be about 2 to 3 times greater than the average power output. The PEP to average power ratio is determined by the shape of the voice waveform. In other words, the ratio is determined by the modulating speech characteristics.

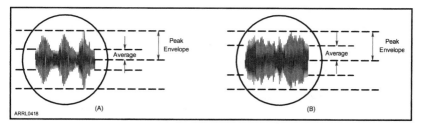

Figure E8A06 — Two RF envelope patterns that show the difference between average and peak levels. In each case, the RF amplitude is plotted as a function of time, as on an oscilloscope display. In (B), the average level is greater, which raises the average output power compared to the peak value.

E8A07   What determines the PEP-to-average power ratio of a single-sideband phone signal?

    A. The frequency of the modulating signal
    B. The characteristics of the modulating signal
    C. The degree of carrier suppression
    D. The amplifier gain

B   See E8A06.

E8A08   Why would a direct or flash conversion analog-to-digital converter be useful for a software defined radio?

    A. Very low power consumption decreases frequency drift
    B. Immunity to out of sequence coding reduces spurious responses
    C. Very high speed allows digitizing high frequencies
    D. All of these choices are correct

C   Flash converters measure the input signal level in a single operation which allows them to sample the input at very high rates compared to other types of converters. The higher rate makes them suitable for sampling wide-bandwidth RF.

E8A09   How many levels can an analog-to-digital converter with 8 bit resolution encode?

    A. 8
    B. 8 multiplied by the gain of the input amplifier
    C. 256 divided by the gain of the input amplifier
    D. 256

D   An analog-to-digital converter can encode 2 to the nth (2^n) different levels, with n being the number of bits in each sample. An 8-bit converter can encode 2 to the 8th (2^8) or 256 different levels.

E8A10   What is the purpose of a low pass filter used in conjunction with a digital-to-analog converter?

    A. Lower the input bandwidth to increase the effective resolution
    B. Improve accuracy by removing out of sequence codes from the input
    C. Remove harmonics from the output caused by the discrete analog levels generated
    D. All of these choices are correct

C   As a digital-to-analog converter jumps between different output levels, the sharp transitions generate high-frequency harmonics and other spurious signals. A low-pass filter removes them from the output signal, leaving only the desired signal.

E8A11 What type of information can be conveyed using digital waveforms?

A. Human speech
B. Video signals
C. Data
D. All of these choices are correct

D Any type of analog signal can be converted into digital information that can then be transmitted as data in a digital waveform.

E8A12 What is an advantage of using digital signals instead of analog signals to convey the same information?

A. Less complex circuitry is required for digital signal generation and detection
B. Digital signals always occupy a narrower bandwidth
C. Digital signals can be regenerated multiple times without error
D. All of these choices are correct

C It is actually more complicated to send and receive signals containing digital data than it is for analog modulation techniques. Nevertheless, the benefits of being able to replicate digital signals exactly for retransmission or storage any number of times outweigh the disadvantages of the additional complexity.

E8A13 Which of these methods is commonly used to convert analog signals to digital signals?

A. Sequential sampling
B. Harmonic regeneration
C. Level shifting
D. Phase reversal

A Measuring a continuous analog waveform at regular intervals is called sequential sampling. Each value in the sequence is a single measurement of the instantaneous amplitude of the waveform at a sampling time. When we make the measurements continually at regular intervals, the result is a sequence of values representing the amplitude of the signal at evenly spaced times.

E8B Modulation and demodulation: modulation methods; modulation index and deviation ratio; frequency and time division multiplexing; orthogonal frequency division multiplexing

E8B01 What is the term for the ratio between the frequency deviation of an RF carrier wave and the modulating frequency of its corresponding FM-phone signal?

A. FM compressibility
B. Quieting index
C. Percentage of modulation
D. Modulation index

D   Modulation index is represented by the symbol band is calculated as $\beta = \Delta f / f_m$. That formula tells you that modulation index is the ratio between the deviation of the frequency-modulated signal (Δf) and the modulating frequency (f_m).

E8B02 How does the modulation index of a phase-modulated emission vary with RF carrier frequency (the modulated frequency)?

A. It increases as the RF carrier frequency increases
B. It decreases as the RF carrier frequency increases
C. It varies with the square root of the RF carrier frequency
D. It does not depend on the RF carrier frequency

D   See E8B01. From the formula for modulation index, you can see that modulation index does not depend on the RF carrier frequency.

E8B03 What is the modulation index of an FM-phone signal having a maximum frequency deviation of 3000 Hz either side of the carrier frequency when the modulating frequency is 1000 Hz?

A. 3
B. 0.3
C. 3000
D. 1000

A   For FM systems being modulated by a tone, the modulation index is given by

$$\beta = \frac{\Delta f}{f_m}$$

where Δf = peak deviation in hertz and f_m = modulating frequency in hertz at any given instant.

Using the values given in the question gives

$$\beta = \frac{3000 \text{ Hz}}{1000 \text{ Hz}} = 3$$

E8B04 What is the modulation index of an FM-phone signal having a maximum carrier deviation of plus or minus 6 kHz when modulated with a 2-kHz modulating frequency?

    A. 6000
    B. 3
    C. 2000
    D. 1/3

B   See E8B03. Using the values given in the question gives

$$\beta = \frac{6000 \text{ Hz}}{2000 \text{ Hz}} = 3$$

E8B05 What is the deviation ratio of an FM-phone signal having a maximum frequency swing of plus-or-minus 5 kHz when the maximum modulation frequency is 3 kHz?

    A. 60
    B. 0.167
    C. 0.6
    D. 1.67

D   Deviation ratio is described by the formula

$$D = \frac{\Delta f}{f_{MAX}}$$

where Δf = peak deviation in hertz and f_{MAX} = maximum modulating frequency in hertz. It measures the ratio of the maximum carrier frequency deviation to the highest audio modulating frequency. The formula for deviation ratio is similar to the one for the modulation index. However, you have to replace the instantaneous modulating frequency, f_m, with the maximum frequency in the modulating signal.

Using the values given in the question gives

$$D = \frac{5000 \text{ Hz}}{3000 \text{ Hz}} = 1.67$$

E8B06 What is the deviation ratio of an FM-phone signal having a maximum frequency swing of plus or minus 7.5 kHz when the maximum modulation frequency is 3.5 kHz?

    A. 2.14
    B. 0.214
    C. 0.47
    D. 47

A   See E8B05. Using the values given in the question gives

$$D = \frac{7.5 \text{ kHz}}{3.5 \text{ kHz}} = 2.14$$

E8B07 Orthogonal Frequency Division Multiplexing is a technique used for which type of amateur communication?

A. High-speed digital modes
B. Extremely low-power contacts
C. EME
D. OFDM signals are not allowed on amateur bands

A Orthogonal Frequency Division Multiplexing (OFDM) consists of multiple carriers separated just enough to avoid interfering with each other at the modulation symbol rate. That non-interfering property is referred to as orthogonality. Because of the multiple carriers, OFDM signals are in the wider class of digital signals.

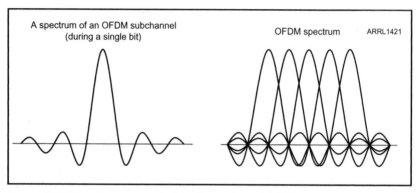

Figure E8B07 — A carrier modulated by a single bit (A) has a spectrum with nulls spaced at regular intervals on either side of the carrier frequency. Multiple carriers placed so their nulls coincide (B) exhibit minimal interference with each other and are called orthogonal. Orthogonal frequency division multiplexing (OFDM) divides the data to be transmitted between many such carriers in wideband digital modes such as the most recent versions of PACTOR.

E8B08 What describes Orthogonal Frequency Division Multiplexing?

A. A frequency modulation technique which uses non-harmonically related frequencies
B. A bandwidth compression technique using Fourier transforms
C. A digital mode for narrow band, slow speed transmissions
D. A digital modulation technique using subcarriers at frequencies chosen to avoid intersymbol interference

D See E8B07.

E8B09 What is meant by deviation ratio?

    A. The ratio of the audio modulating frequency to the center carrier frequency

    B. The ratio of the maximum carrier frequency deviation to the highest audio modulating frequency

    C. The ratio of the carrier center frequency to the audio modulating frequency

    D. The ratio of the highest audio modulating frequency to the average audio modulating frequency

B   See E8B05.

E8B10 What describes frequency division multiplexing?

    A. The transmitted signal jumps from band to band at a predetermined rate

    B. Two or more information streams are merged into a baseband, which then modulates the transmitter

    C. The transmitted signal is divided into packets of information

    D. Two or more information streams are merged into a digital combiner, which then pulse position modulates the transmitter

B   Multiplexing is defined as combining several streams of information into a single stream. Of the various schemes for combining the streams, frequency division multiplexing (FDM) is often used to combine analog information streams for analog modulation. The usual method is to shift the information streams to different frequencies, add them together to create a baseband signal, then modulate a carrier with the baseband signal.

E8B11 What is digital time division multiplexing?

    A. Two or more data streams are assigned to discrete sub-carriers on an FM transmitter

    B. Two or more signals are arranged to share discrete time slots of a data transmission

    C. Two or more data streams share the same channel by transmitting time of transmission as the sub-carrier

    D. Two or more signals are quadrature modulated to increase bandwidth efficiency

B   In time division multiplexing (TDM), the information streams are divided into time slices. The individual slices are then interleaved into a single information stream. This process works best for digital information.

E8C Digital signals: digital communication modes; information rate vs bandwidth; error correction

E8C01   How is Forward Error Correction implemented?

A. By the receiving station repeating each block of three data characters

B. By transmitting a special algorithm to the receiving station along with the data characters

C. By transmitting extra data that may be used to detect and correct transmission errors

D. By varying the frequency shift of the transmitted signal according to a predefined algorithm

C   This is a general question about Forward Error Correction systems. These systems do not require receiving stations to transmit an acknowledgement to the sending station. FEC is implemented by transmitting extra data that may be used to detect and correct transmission errors.

E8C02   What is the definition of symbol rate in a digital transmission?

A. The number of control characters in a message packet

B. The duration of each bit in a message sent over the air

C. The rate at which the waveform of a transmitted signal changes to convey information

D. The number of characters carried per second by the station-to-station link

C   Digital data symbols are encoded as specific changes in the transmitted signal. Thus the symbol rate is the rate at which the transmitted signal changes to carry information.

E8C03   When performing phase shift keying, why is it advantageous to shift phase precisely at the zero crossing of the RF carrier?

A. This results in the least possible transmitted bandwidth for the particular mode

B. It is easier to demodulate with a conventional, non-synchronous detector

C. It improves carrier suppression

D. All of these choices are correct

A   By changing the signal's phase when the RF carrier signal waveform is at zero, the change to the transmitted signal is minimized. This requires the least amount of bandwidth on the air.

E8C04 What technique is used to minimize the bandwidth requirements of a PSK31 signal?

A. Zero-sum character encoding
B. Reed-Solomon character encoding
C. Use of sinusoidal data pulses
D. Use of trapezoidal data pulses

C If a modulating signal has sharp edges, such as a for a digital data stream, then the resulting modulated signal must have enough bandwidth to reproduce the high-frequency components of the modulating signal that create the sharp edges. By using a more rounded modulating signal, less bandwidth is required for the modulated signal. PSK31 uses sinusoidally shaped pulses as its method of minimizing bandwidth requirements.

E8C05 What is the necessary bandwidth of a 13-WPM International Morse code transmission?

A. Approximately 13 Hz
B. Approximately 26 Hz
C. Approximately 52 Hz
D. Approximately 104 Hz

C You can calculate the necessary bandwidth of a CW signal with this formula

$$BW = B \times K$$

where BW = the necessary bandwidth of the signal in Hz, B = the data rate in bauds of the transmission; and K = a factor relating to shape of the keying envelope.

Divide WPM by 1.2 to convert to bauds. K is typically between 3 (soft keying) and 5 (hard keying). A typical value for K is 4.8. The necessary bandwidth for a CW signal then becomes

$$BW = \frac{WPM}{1.2} \times K = \frac{13}{1.2} \times 4.8 = 52 \text{ Hz}$$

If you look at the formula you can see that with a K of 4.8, the necessary bandwidth of a CW signal in Hz is the speed in WPM times 4.

E8C06   What is the necessary bandwidth of a 170-hertz shift, 300-baud ASCII transmission?

    A. 0.1 Hz
    B. 0.3 kHz
    C. 0.5 kHz
    D. 1.0 kHz

C   An AFSK data signal is generated by injecting two audio tones, separated by the correct shift into the microphone input of an SSB transmitter. The necessary bandwidth for this type of data transmission is

$$BW = (K \times Shift) + B$$

where BW = the necessary bandwidth in hertz; K = a constant that for Amateur Radio you can assume to be 1.2; Shift = frequency shift in hertz; and B = data rate in bauds.

Using the values from the question you get

$$BW = (K \times Shift) + B = (1.2 \times 170) + 300 = 504 \text{ Hz}$$

This is approximately 0.5 kHz.

E8C07   What is the necessary bandwidth of a 4800-Hz frequency shift, 9600-baud ASCII FM transmission?

    A. 15.36 kHz
    B. 9.6 kHz
    C. 4.8 kHz
    D. 5.76 kHz

A   See E8C06. Using the values from the question you get

$$BW = (K \times Shift) + B = (1.2 \times 4800) + 9600 = 15.36 \text{ kHz}$$

E8C08   How does ARQ accomplish error correction?

    A. Special binary codes provide automatic correction
    B. Special polynomial codes provide automatic correction
    C. If errors are detected, redundant data is substituted
    D. If errors are detected, a retransmission is requested

D   ARQ stands for Automatic Repeat reQuest. When an error is detected in received data, the ARQ receive system sends a NAK (Not Acknowledge) message back to the transmitting system so that the data is sent again.

E8C09   What is the name of a digital code where each preceding or following character changes by only one bit?

A. Binary Coded Decimal Code
B. Extended Binary Coded Decimal Interchange Code
C. Excess 3 code
D. Gray code

D   Gray code is used when it is useful to minimize the amount of change between symbols.

E8C10   What is an advantage of Gray code in digital communications where symbols are transmitted as multiple bits?

A. It increases security
B. It has more possible states than simple binary
C. It has more resolution than simple binary
D. It facilitates error detection

D   Because in Gray code only one bit changes between successive symbols, that property can be used to detect errors.

E8C11   What is the relationship between symbol rate and baud?

A. They are the same
B. Baud is twice the symbol rate
C. Symbol rate is only used for packet-based modes
D. Baud is only used for RTTY

A   Baud is the unit for symbol rate. See also E8C02.

E8D   Keying defects and overmodulation of digital signals; digital codes; spread spectrum

E8D01   Why are received spread spectrum signals resistant to interference?

A. Signals not using the spread spectrum algorithm are suppressed in the receiver
B. The high power used by a spread spectrum transmitter keeps its signal from being easily overpowered
C. The receiver is always equipped with a digital blanker
D. If interference is detected by the receiver it will signal the transmitter to change frequencies

A   Because the spread-spectrum signal changes frequencies rapidly, interference or noise on a single frequency affects it only briefly. To interfere with a spread-spectrum signal, interference would have to follow the changing frequency exactly or cover a significant fraction of the band it occupies.

E8D02 What spread spectrum communications technique uses a high speed binary bit stream to shift the phase of an RF carrier?

A. Frequency hopping
B. Direct sequence
C. Binary phase-shift keying
D. Phase compandored spread spectrum

B Direct sequence spread spectrum uses a very fast binary bit stream to shift the phase of an RF carrier.

E8D03 How does the spread spectrum technique of frequency hopping work?

A. If interference is detected by the receiver it will signal the transmitter to change frequencies
B. If interference is detected by the receiver it will signal the transmitter to wait until the frequency is clear
C. A pseudo-random binary bit stream is used to shift the phase of an RF carrier very rapidly in a particular sequence
D. The frequency of the transmitted signal is changed very rapidly according to a particular sequence also used by the receiving station

D Frequency hopping (FH) is a form of spread spectrum in which the center frequency of a conventional carrier is altered many times per second in accordance with a list of frequency channels. The same frequency list is used by the receiving station, so it hops in sync with the transmitter.

E8D04 What is the primary effect of extremely short rise or fall time on a CW signal?

A. More difficult to copy
B. The generation or RF harmonics
C. The generation of key clicks
D. Limits data speed

C The sharp rising and falling edges of the keying waveform require high-frequency components that modulate the transmitter and cause interference to signals on adjacent frequencies.

E8D05 What is the most common method of reducing key clicks?

A. Increase keying waveform rise and fall times
B. Low-pass filters at the transmitter output
C. Reduce keying waveform rise and fall times
D. High-pass filters at the transmitter output

A Many transceivers feature menu or circuit changes to adjust keying rise and fall time. (See also E8D04.)

E8D06 Which of the following indicates likely overmodulation of an AFSK signal such as PSK or MFSK?

A. High reflected power
B. Strong ALC action
C. Harmonics on higher bands
D. Rapid signal fading

B The ALC (automatic level control) system adjusts audio and drive level to keep from overdriving the final amplifier stages. This distorts the input signal to some degree and strong ALC action indicates the input audio level to the transmitter is too high.

E8D07 What is a common cause of overmodulation of AFSK signals?

A. Excessive numbers of retries
B. Ground loops
C. Bit errors in the modem
D. Excess transmit audio levels

D Just as with a speech signal, excessive audio level of any AFSK signal at the transmitter input will cause overmodulation.

E8D08 What parameter might indicate that excessively high input levels are causing distortion in an AFSK signal?

A. Signal to noise ratio
B. Baud rate
C. Repeat Request Rate (RRR)
D. Intermodulation Distortion (IMD)

D Intermodulation distortion is automatically measured by many digital mode software packages and the information can be useful to the transmitting station in order to keep the transmissions clean and free of spurious signals and sidebands.

E8D09 What is considered a good minimum IMD level for an idling PSK signal?

A. +10 dB
B. +15 dB
C. –20 dB
D. –30 dB

D This value means the spurious emissions from the PSK transmitter are 1000 times weaker than the main PSK signal. Lower numbers indicate a cleaner signal.

E8D10 What are some of the differences between the Baudot digital code and ASCII?

    A. Baudot uses 4 data bits per character, ASCII uses 7 or 8; Baudot uses 1 character as a letters/figures shift code, ASCII has no letters/figures code

    B. Baudot uses 5 data bits per character, ASCII uses 7 or 8; Baudot uses 2 characters as letters/figures shift codes, ASCII has no letters/figures shift code

    C. Baudot uses 6 data bits per character, ASCII uses 7 or 8; Baudot has no letters/figures shift code, ASCII uses 2 letters/figures shift codes

    D. Baudot uses 7 data bits per character, ASCII uses 8; Baudot has no letters/figures shift code, ASCII uses 2 letters/figures shift codes

B   Baudot code was invented in the early days of teleprinters and only has bits. ASCII was invented after World War II and has enough bits to represent all of the letters (upper and lower case), numerals, and punctuation, as well as various control characters, without requiring the use of a shift character to select between alphabetic and numeric characters sets.

E8D11 What is one advantage of using ASCII code for data communications?

    A. It includes built in error correction features

    B. It contains fewer information bits per character than any other code

    C. It is possible to transmit both upper and lower case text

    D. It uses one character as a shift code to send numeric and special characters

C   See E8D10.

E8D12 What is the advantage of including a parity bit with an ASCII character stream?

    A. Faster transmission rate

    B. The signal can overpower interfering signals

    C. Foreign language characters can be sent

    D. Some types of errors can be detected

D   Parity counts the number of 1 bits in a string or character. A parity bit can be set to indicate whether the number of bits was even or odd. This allows single-bit errors in the string to be detected.

Antennas and Transmission Lines

There will be eight questions on your Extra class examination from the Antennas and Transmission Lines subelement. These eight questions will be taken from the eight groups of questions labeled E9A through E9H.

E9A   Basic antenna parameters: radiation resistance, gain, beamwidth, efficiency, beamwidth; effective radiated power, polarization

E9A01   What describes an isotropic antenna?

A. A grounded antenna used to measure earth conductivity
B. A horizontally polarized antenna used to compare Yagi antennas
C. A theoretical antenna used as a reference for antenna gain
D. A spacecraft antenna used to direct signals toward the earth

C   You'll never encounter an isotropic radiator because it is a theoretical (mathematical) concept, useful only for comparing antenna performance. It is a point-source radiator located in space that exhibits no directivity in any direction. In other words, it radiates equally in all directions and its radiation pattern is perfectly spherical.

E9A02   What antenna has no gain in any direction?

A. Quarter-wave vertical
B. Yagi
C. Half-wave dipole
D. Isotropic antenna

D   By definition, an isotropic radiator transmits uniformly in all directions. Gain is created by a departure from this uniform pattern.

E9A03 Why would one need to know the feed point impedance of an antenna?

A. To match impedances in order to minimize standing wave ratio on the transmission line
B. To measure the near-field radiation density from a transmitting antenna
C. To calculate the front-to-side ratio of the antenna
D. To calculate the front-to-back ratio of the antenna

A Causing as much of the available RF energy to be radiated as possible is accomplished, in part, by matching the antenna impedance to the feed line. To match the impedances, you must know the antenna's feed point impedance or create an adjustable matching system.

E9A04 Which of the following factors may affect the feed point impedance of an antenna?

A. Transmission-line length
B. Antenna height, conductor length/diameter ratio and location of nearby conductive objects
C. The settings of an antenna tuner at the transmitter
D. Sunspot activity and time of day

B An antenna's location with respect to nearby objects — especially the Earth — helps determine the radiation resistance. So does the conductors' length-to-diameter (L/D) ratio.

E9A05 What is included in the total resistance of an antenna system?

A. Radiation resistance plus space impedance
B. Radiation resistance plus transmission resistance
C. Transmission-line resistance plus radiation resistance
D. Radiation resistance plus ohmic resistance

D Radiation resistance is an assumed resistance that represents the power actually radiated from the antenna. Real, or ohmic, resistance in the system dissipates energy as heat. The total resistance of an antenna system is the sum of the radiation resistance plus ohmic resistance.

E9A06 How does the beamwidth of an antenna vary as the gain is increased?

A. It increases geometrically
B. It increases arithmetically
C. It is essentially unaffected
D. It decreases

D The higher the gain, the more directive the antenna pattern. In other words, the beamwidth will decrease as gain increases.

E9A07   **What is meant by antenna gain?**

    A. The ratio of the radiated signal strength of an antenna in the direction of maximum radiation to that of a reference antenna

    B. The ratio of the signal in the forward direction to that in the opposite direction

    C. The ratio of the amount of power radiated by an antenna compared to the transmitter output power

    D. The final amplifier gain minus the transmission line losses

A   Gain measures the directivity of an antenna compared to a reference antenna. For example, the peak gain of a dipole is 2.15 dB greater than that of an isotropic antenna.

E9A08   **What is meant by antenna bandwidth?**

    A. Antenna length divided by the number of elements

    B. The frequency range over which an antenna satisfies a performance requirement

    C. The angle between the half-power radiation points

    D. The angle formed between two imaginary lines drawn through the element ends

B   Bandwidth is the frequency range over which an antenna can be expected to meet some specified level of performance, such as an SWR value. For example, the range over which the antenna exhibits an SWR of 2:1 or less. Antennas also have gain bandwidth and front-to-back bandwidth.

E9A09   **How is antenna efficiency calculated?**

    A. (radiation resistance / transmission resistance) x 100 percent

    B. (radiation resistance / total resistance) x 100 percent

    C. (total resistance / radiation resistance) x 100 percent

    D. (effective radiated power / transmitter output) x 100 percent

B   Calculate the efficiency of an antenna by dividing the radiation resistance by the total resistance. Multiply by 100 to get the answer in percent. Total resistance is radiation resistance plus ohmic resistance.

E9A10   **Which of the following choices is a way to improve the efficiency of a ground-mounted quarter-wave vertical antenna?**

    A. Install a good radial system

    B. Isolate the coax shield from ground

    C. Shorten the radiating element

    D. Reduce the diameter of the radiating element

A   To significantly improve antenna efficiency, lower the antenna system's ohmic resistance. In the case of a typical HF grounded vertical antenna, losses in the ground system are the chief source of ohmic resistance. Installing a good ground radial system can lower this resistance.

E9A11 Which of the following factors determines ground losses for a ground-mounted vertical antenna operating in the 3 MHz to 30 MHz range?

A. The standing wave ratio
B. Distance from the transmitter
C. Soil conductivity
D. Take-off angle

C Ground-mounted ¼-wavelength vertical antennas depend on a ground system to create an electrical image that creates the missing half of the ½-wavelength dipole. If the ground system consists largely of soil, losses are determined by the soil's conductivity. This is why installing a ground system of radial wires can greatly reduce ground system losses.

E9A12 How much gain does an antenna have compared to a 1/2-wavelength dipole when it has 6 dB gain over an isotropic antenna?

A. 3.85 dB
B. 6.0 dB
C. 8.15 dB
D. 2.79 dB

A The ½-wavelength dipole has a gain of 2.15 dB relative to an isotropic antenna. The gain relative to the dipole would then be 6 dB – 2.15 dB = 3.85 dB.

E9A13 How much gain does an antenna have compared to a ½-wavelength dipole when it has 12 dB gain over an isotropic antenna?

A. 6.17 dB
B. 9.85 dB
C. 12.5 dB
D. 14.15 dB

B Again, the ½-wavelength dipole has a gain of 2.15 dB relative to an isotropic antenna. The gain relative to the dipole would then be 12 dB – 2.15 dB = 9.85 dB.

E9A14 What is meant by the radiation resistance of an antenna?

A. The combined losses of the antenna elements and feed line
B. The specific impedance of the antenna
C. The value of a resistance that would dissipate the same amount of power as that radiated from an antenna
D. The resistance in the atmosphere that an antenna must overcome to be able to radiate a signal

C Radiation resistance is the equivalent resistance that would dissipate the same amount of power as that radiated from an antenna.

E9A15 What is the effective radiated power relative to a dipole of a repeater station with 150 watts transmitter power output, 2 dB feed line loss, 2.2 dB duplexer loss, and 7 dBd antenna gain?

    A. 1977 watts
    B. 78.7 watts
    C. 420 watts
    D. 286 watts

D   This is the first of a series of questions related to dB and effective radiated power (ERP). Be careful to check whether the question uses a dipole (ERP) or an isotropic radiator (EIRP) as the reference antenna. System gains and losses are usually expressed in dB, because it makes computations easier.

$$ERP = TPO \times \log^{-1}\left(\frac{\text{system loss or gain}}{10}\right)$$

where TPO = transmitter power output and (system loss or gain) = the sum of gains and losses starting at the transmitter and including antenna gain.

$$ERP = 150 \times \log^{-1}\left(\frac{-2 - 2.2 + 7}{10}\right) = 150 \times \log^{-1}(0.28) = 286 \text{ W}$$

E9A16 What is the effective radiated power relative to a dipole of a repeater station with 200 watts transmitter power output, 4 dB feed line loss, 3.2 dB duplexer loss, 0.8 dB circulator loss, and 10 dBd antenna gain?

    A. 317 watts
    B. 2000 watts
    C. 126 watts
    D. 300 watts

A   See E9A15.

$$ERP = 200 \times \log^{-1}\left(\frac{-4 - 3.2 - 0.8 + 10}{10}\right) = 200 \times \log^{-1}(0.2) = 317 \text{ W}$$

E9A17 What is the effective isotropic radiated power of a repeater station with 200 watts transmitter power output, 2 dB feed line loss, 2.8 dB duplexer loss, 1.2 dB circulator loss and 7 dBi antenna gain?

    A. 159 watts
    B. 252 watts
    C. 632 watts
    D. 63.2 watts

B   See E9A15.

$$ERP = 200 \times \log^{-1}\left(\frac{-2 - 2.8 - 1.2 + 7}{10}\right) = 200 \times \log^{-1}(0.1) = 252 \text{ W}$$

E9A18 What term describes station output, taking into account all gains and losses?

    A. Power factor
    B. Half-power bandwidth
    C. Effective radiated power
    D. Apparent power

C See E9A15. The intent of using effective radiated power (ERP) is to be able to compare different stations based on their transmitter output power and an equivalent antenna, the dipole. To do so, all of the antenna system gains and losses are combined into one compensating factor.

E9B Antenna patterns: E and H plane patterns; gain as a function of pattern; antenna design

E9B01 In the antenna radiation pattern shown in Figure E9-1, what is the 3 dB beamwidth?

    A. 75 degrees
    B. 50 degrees
    C. 25 degrees
    D. 30 degrees

B By looking at the figure, the strength of the radiated signal drops 3 dB at about ±25 degrees from the peak of the main lobe. That makes the beamwidth about 50 degrees.

E9B02 In the antenna radiation pattern shown in Figure E9-1, what is the front-to-back ratio?

    A. 36 dB
    B. 18 dB
    C. 24 dB
    D. 14 dB

B The antenna gain at 0 degrees is 0 dB while the gain at 180 degrees is about halfway between −12 and −24 dB on this scale. You should read this as a ratio of 18 dB.

E9B03 In the antenna radiation pattern shown in Figure E9-1, what is the front-to-side ratio?

    A. 12 dB
    B. 14 dB
    C. 18 dB
    D. 24 dB

B The antenna gain at 0 degrees is 0 dB while the gain at 90 degrees and 270 degrees is a bit less than −12 dB and can be estimated as a ratio of 14 dB.

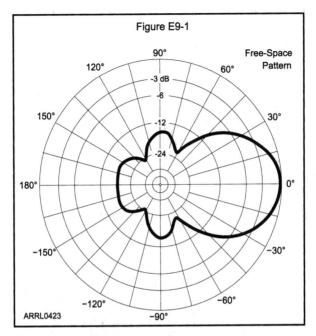

Figure E9-1

90°
120°   60°
          -3 dB
          -6
150°   -12   30°
          -24
180°   0°

-150°   -30°

-120°   -60°
ARRL0423   -90°

Free-Space Pattern

Figure E9-1 — Use this graph for E9B01 through E9B03.

E9B04 What may occur when a directional antenna is operated at different frequencies within the band for which it was designed?

A. Feed point impedance may become negative
B. The E-field and H-field patterns may reverse
C. Element spacing limits could be exceeded
D. The gain may change depending on frequency

D   Antenna gain changes with frequency because the electrical length and spacing of its elements change. In antennas that are sensitive to small changes in element characteristics, such as Yagi antennas, gain and other parameters can change significantly across a frequency band.

E9B05 What type of antenna pattern over real ground is shown in Figure E9-2?

A. Elevation
B. Azimuth
C. Radiation resistance
D. Polarization

A The antenna pattern shown in Figure E9-2 is an elevation pattern. An elevation pattern over real ground usually shows only half a circle. Any radiation that would have gone down into the ground is reflected back into space above the Earth, and that energy is added to energy radiated directly from the antenna, creating the lobes and nulls of the elevation pattern.

E9B06 What is the elevation angle of peak response in the antenna radiation pattern shown in Figure E9-2?

A. 45 degrees
B. 75 degrees
C. 7.5 degrees
D. 25 degrees

C By looking at the radiation pattern, you can see the largest lobe is centered at 7.5 degrees above the horizon. This is the peak response.

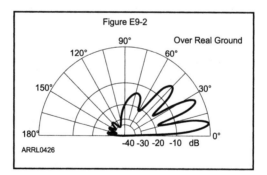

Figure E9-2 — Use this graph for E9B05, E9B06, E9B15, and E9B16.

E9B07 How does the total amount of radiation emitted by a directional gain antenna compare with the total amount of radiation emitted from an isotropic antenna, assuming each is driven by the same amount of power?

A. The total amount of radiation from the directional antenna is increased by the gain of the antenna

B. The total amount of radiation from the directional antenna is stronger by its front to back ratio

C. They are the same

D. The radiation from the isotropic antenna is 2.15 dB stronger than that from the directional antenna

C The total amount of power radiated is the same for both antennas (ignoring heat losses from the real or ohmic resistance) but the directional antenna focuses it in one or more directions. Gain only means that the antenna has directivity, not that additional power is created.

E9B08 How can the approximate beamwidth in a given plane of a directional antenna be determined?

A. Note the two points where the signal strength of the antenna is 3 dB less than maximum and compute the angular difference

B. Measure the ratio of the signal strengths of the radiated power lobes from the front and rear of the antenna

C. Draw two imaginary lines through the ends of the elements and measure the angle between the lines

D. Measure the ratio of the signal strengths of the radiated power lobes from the front and side of the antenna

A The beamwidth is typically defined in terms of the half-power points. These are the points on either side and closest to the main lobe where the gain 3 dB below the peak gain. The beamwidth is the angular difference between these two points.

E9B09 What type of computer program technique is commonly used for modeling antennas?

A. Graphical analysis

B. Method of Moments

C. Mutual impedance analysis

D. Calculus differentiation with respect to physical properties

B See E9B10. Antenna modeling software uses the Method of Moments to analyze antenna performance. This analysis technique divides an antenna into segments, computes the current in each segment, and sums the radiation resulting from the currents in all segments.

E9B10 What is the principle of a Method of Moments analysis?

A. A wire is modeled as a series of segments, each having a uniform value of current
B. A wire is modeled as a single sine-wave current generator
C. A wire is modeled as a series of points, each having a distinct location in space
D. A wire is modeled as a series of segments, each having a distinct value of voltage across it

A The Method of Moments analysis technique divides wires into segments and assigns an appropriate current value to each segment. The current in and radiation from each segment is called a moment.

E9B11 What is a disadvantage of decreasing the number of wire segments in an antenna model below the guideline of 10 segments per half-wavelength?

A. Ground conductivity will not be accurately modeled
B. The resulting design will favor radiation of harmonic energy
C. The computed feed point impedance may be incorrect
D. The antenna will become mechanically unstable

C By reducing the number of the segments so that they become longer, the assumption that current is constant across the segment becomes less valid. This degrades the accuracy of the final calculations.

E9B12 What is the far field of an antenna?

A. The region of the ionosphere where radiated power is not refracted
B. The region where radiated power dissipates over a specified time period
C. The region where radiated field strengths are obstructed by objects of reflection
D. The region where the shape of the antenna pattern is independent of distance

D Close to the antenna, the antenna pattern has not fully formed. When the waves have travelled far enough to reach their final relationship in amplitude and phase, forming the actual radiation pattern, that is considered the antenna's far field.

E9B13   What does the abbreviation NEC stand for when applied to antenna modeling programs?

A. Next Element Comparison
B. Numerical Electromagnetics Code
C. National Electrical Code
D. Numeric Electrical Computation

B   The Numerical Electromagnetics Code (NEC) was developed and validated as publically funded research, then placed in the public domain. Many antenna programs make use of its fundamental algorithms. There are several versions, with the latest being NEC-4.

E9B14   What type of information can be obtained by submitting the details of a proposed new antenna to a modeling program?

A. SWR vs frequency charts
B. Polar plots of the far-field elevation and azimuth patterns
C. Antenna gain
D. All of these choices are correct

D   Antenna modeling programs can calculate all of these parameters and many more. The ability for individual amateurs to have sophisticated modeling programs has created many opportunities for amateur antenna experimentation.

E9B15   What is the front-to-back ratio of the radiation pattern shown in Figure E9-2?

A. 15 dB
B. 28 dB
C. 3 dB
D. 24 dB

B   The back lobes rise to just above the –30 dB circle and the main lobe reaches the 0 dB circle. This means you have a front-to-back ratio of about 28 dB.

E9B16   How many elevation lobes appear in the forward direction of the antenna radiation pattern shown in Figure E9-2?

A. 4
B. 3
C. 1
D. 7

A   The forward direction includes those lobes between 0 degrees and 90 degrees and there are four lobes in that range.

E9C   Wire and phased array antennas: rhombic antennas; effects of ground reflections; take-off angles; practical wire antennas: Zepps, OCFD, loops

E9C01   What is the radiation pattern of two ¼ wavelength vertical antennas spaced ½ wavelength apart and fed 180 degrees out of phase?

A. Cardioid
B. Omni-directional
C. A figure-8 broadside to the axis of the array
D. A figure-8 oriented along the axis of the array

D   Phased-array antennas create their radiation patterns by controlling the phase of the feed point current and the spacing of the antennas. The resulting phase differences of the radiated signal from each element create the radiation pattern's lobes and nulls through constructive and destructive interference, respectively. Figure E9C01 shows horizontal directive patterns of two phased verticals, spaced and phased as indicated. The radiation pattern of two ¼-wavelength vertical antennas spaced ½ wavelength apart and fed 180 degrees out of phase is a figure-8 end-fire in line with the antennas.

E9C02   What is the radiation pattern of two ¼ wavelength vertical antennas spaced ¼ wavelength apart and fed 90 degrees out of phase?

A. Cardioid
B. A figure-8 end-fire along the axis of the array
C. A figure-8 broadside to the axis of the array
D. Omni-directional

A   See E9C01.

E9C03   What is the radiation pattern of two ¼-wavelength vertical antennas spaced ½ wavelength apart and fed in phase?

A. Omni-directional
B. Cardioid
C. A Figure-8 broadside to the axis of the array
D. A Figure-8 end-fire along the axis of the array

C   See E9C01.

Figure E9C01 — Horizontal directive patterns of two phased verticals, spaced and phased as indicated. In these plots, you are looking directly onto the axis of the antennas (the antennas are perpendicular to the page) and the pair of antennas are arranged along a line running vertically ("North-South") through the plot. The uppermost (northern) element is lagging if there is a phase difference between the two antennas. The antennas are identical with the same magnitude of current in each. Gain is stated with respect to a single vertical.

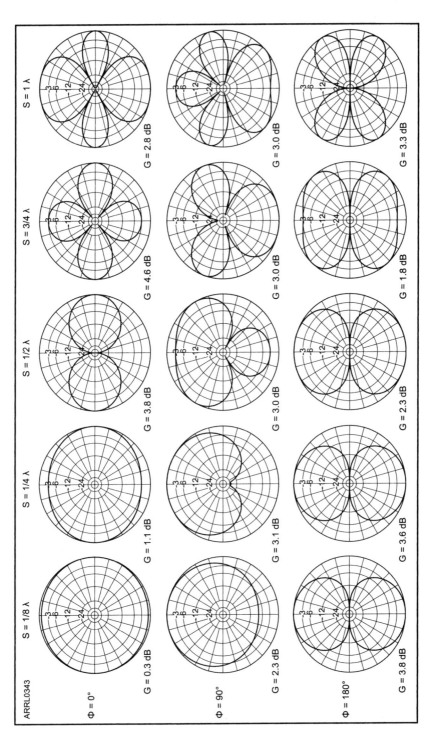

E9C04 What happens to the radiation pattern of an unterminated long wire antenna as the wire length is increased?

A. The lobes become more perpendicular the wire
B. The lobes align more in the direction of the wire
C. The vertical angle increases
D. The front-to-back ratio decreases

B For half-wave dipoles, radiation is strongest in one main lobe broadside to the antenna. As the antenna becomes longer and longer, the main lobe breaks up into an increasing number of lobes that become aligned more closely with the antenna.

E9C05 What is an OCFD antenna?

A. A dipole fed approximately ⅓ the way from one end with a 4:1 balun to provide multiband operation
B. A remotely tunable dipole antenna using orthogonally controlled frequency diversity
C. An eight band dipole antenna using octophase filters
D. A multiband dipole antenna using one-way circular polarization for frequency diversity

A OCFD stands for Off-Center-Fed Dipole. By feeding the antenna away from the low-impedance point at the center of the dipole on its fundamental frequency, a more consistent impedance can be obtained on multiple bands. An impedance transformer at the feed point then matches the antenna to coaxial feed line on multiple bands.

E9C06 What is the effect of a terminating resistor on a rhombic antenna?

A. It reflects the standing waves on the antenna elements back to the transmitter
B. It changes the radiation pattern from bidirectional to unidirectional
C. It changes the radiation pattern from horizontal to vertical polarization
D. It decreases the ground loss

B The main effect of a terminating resistor on a rhombic antenna is to change the radiation pattern from essentially bidirectional to essentially unidirectional by absorbing power that would otherwise be reflected back toward the feed point and radiate in the opposite direction.

E9C07   What is the approximate feed point impedance at the center of a two-wire folded dipole antenna?

    A. 300 ohms
    B. 72 ohms
    C. 50 ohms
    D. 450 ohms

A   The input impedance of a dipole is approximately 73 ohms. In a folded dipole antenna, the current is divided between the two parallel wires. That means that the current at the feed point of a folded dipole is half of what it is in a dipole. From the power formula $P = I^2 \times Z$, you can see that with the same power and half the current, you'll also have four times the impedance. In this case approximately 300 ohms.

E9C08   What is a folded dipole antenna?

    A. A dipole one-quarter wavelength long
    B. A type of ground-plane antenna
    C. A dipole constructed from one wavelength of wire forming a very thin loop
    D. A dipole configured to provide forward gain

C   A folded dipole antenna is a wire antenna that consists of one full wavelength of wire — unlike the ½ wavelength of wire in a regular dipole — in a very thin loop. It is fed at the center of one of the parallel wires as shown in Figure E9C08. The parallel wires act to raise the feed point impedance, which is useful in some applications.

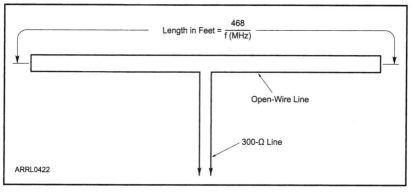

Figure E9C08 — A half-wave folded dipole consists of one full wavelength of wire in a very thin loop.

E9C09 What is a G5RV antenna?

A. A multi-band dipole antenna fed with coax and a balun through a selected length of open wire transmission line
B. A multi-band trap antenna
C. A phased array antenna consisting of multiple loops
D. A wide band dipole using shorted coaxial cable for the radiating elements and fed with a 4:1 balun

A The G5RV-style antenna uses a section of open-wire feed line to transform the impedance of the dipole section to match coaxial cable on several bands. A choke balun is used at the transition from open-wire to coaxial feed line to stabilize feed point impedance and preserve antenna system balance by isolating the outer surface of the coax feed line shield.

E9C10 Which of the following describes a Zepp antenna?

A. A dipole constructed from zip cord
B. An end fed dipole antenna
C. An omni-directional antenna commonly used for satellite communications
D. A vertical array capable is quickly changing the direction of maximum radiation by changing phasing lines

B The Zepp — short for Zeppelin — is a half-wavelength dipole fed at one end through open-wire feed line. The antenna was originally used aboard the Zeppelin airships where it could only be supported at one end.

E9C11 How is the far-field elevation pattern of a vertically polarized antenna affected by being mounted over seawater versus rocky ground?

A. The low-angle radiation decreases
B. The high-angle radiation increases
C. Both the high-angle and low-angle radiation decrease
D. The low-angle radiation increases

D Seawater has excellent conductivity and rocky ground is a poor conductor. If you mount a vertically polarized antenna over seawater, it will have the effect of increasing the low-angle radiation as compared to a similar antenna over rocky ground.

E9C12 **Which of the following describes an extended double Zepp antenna?**

    A. A wideband vertical antenna constructed from precisely tapered aluminum tubing

    B. A portable antenna erected using two push support poles

    C. A center fed 1.25 wavelength antenna (two ⅝ wave elements in phase)

    D. An end fed folded dipole antenna

C   Each half of the antenna is an extended Zepp approximately ⅝-wavelength long. (See also E9C10.) A section of open-wire feed line transforms the feed point to match coaxial feed line.

E9C13 **What is the main effect of placing a vertical antenna over an imperfect ground?**

    A. It causes increased SWR

    B. It changes the impedance angle of the matching network

    C. It reduces low-angle radiation

    D. It reduces losses in the radiating portion of the antenna

C   The radiation pattern of an antenna over real ground is always affected by the electrical conductivity and dielectric constant of the soil. This is especially true of the low-elevation-angle far-field pattern of a vertically polarized antenna. The low-angle radiation pattern from a vertically polarized antenna mounted over seawater will be much stronger than for a similar antenna mounted over rocky soil, for example. The less the loss in the ground around the antenna, the more radiation will occur at low vertical angles.

E9C14 **How does the performance of a horizontally polarized antenna mounted on the side of a hill compare with the same antenna mounted on flat ground?**

    A. The main lobe takeoff angle increases in the downhill direction

    B. The main lobe takeoff angle decreases in the downhill direction

    C. The horizontal beamwidth decreases in the downhill direction

    D. The horizontal beamwidth increases in the uphill direction

B   Ground reflections in the downward-sloping direction reinforce the main lobe of the antenna at a lower vertical angle which is better for long-distance communication.

E9C15 How does the radiation pattern of a horizontally polarized 3-element beam antenna vary with its height above ground?

    A. The main lobe takeoff angle increases with increasing height
    B. The main lobe takeoff angle decreases with increasing height
    C. The horizontal beamwidth increases with height
    D. The horizontal beamwidth decreases with height

B   In general, the radiation takeoff angle from a Yagi antenna with horizontally mounted elements decreases as the antenna height increases above flat ground. So, if you raise the height of your antenna, the takeoff angle will decrease.

E9D   Directional antennas: gain; Yagi antennas; losses; SWR bandwidth; antenna efficiency; shortened and mobile antennas; RF grounding

E9D01 How does the gain of an ideal parabolic dish antenna change when the operating frequency is doubled?

    A. Gain does not change
    B. Gain is multiplied by 0.707
    C. Gain increases by 6 dB
    D. Gain increases by 3 dB

C   For a parabolic dish, the gain is proportional to the square of the frequency so if you double the frequency, you will raise the gain by a factor of four. In dB units, a factor of 4 is a gain of 6 dB.

E9D02 How can linearly polarized Yagi antennas be used to produce circular polarization?

    A. Stack two Yagis fed 90 degrees out of phase to form an array with the respective elements in parallel planes
    B. Stack two Yagis fed in phase to form an array with the respective elements in parallel planes
    C. Arrange two Yagis perpendicular to each other with the driven elements at the same point on the boom fed 90 degrees out of phase
    D. Arrange two Yagis collinear to each other, with the driven elements fed 180 degrees out of phase

C   Two Yagi antennas built on the same boom, with elements placed perpendicular to each other, form the basis of a circularly polarized antenna as shown in Figure E9D02. The driven elements are located at the same position along the boom, so they lie on the same plane, which is perpendicular to the boom. The driven elements are then fed 90 degrees out of phase.

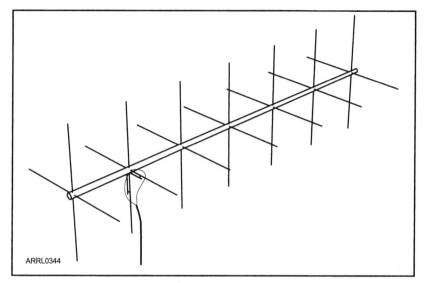

Figure E9D02 — A circularly polarized antenna built from two Yagis on the same boom.

E9D03   Where should a high-Q loading coil be placed to minimize losses in a shortened vertical antenna?

A. Near the center of the vertical radiator
B. As low as possible on the vertical radiator
C. As close to the transmitter as possible
D. At a voltage node

A   As you move a loading coil higher on the vertical, it requires more inductance to bring the antenna to resonance. More inductance results in more ohmic losses. However, as the coil moves up, the radiation resistance of the antenna goes up, which results in higher efficiency. You'll find that a loading coil placed near the center of the vertical balances the two effects for the most effective performance from the antenna.

E9D04   Why should an HF mobile antenna loading coil have a high ratio of reactance to resistance?

A. To swamp out harmonics
B. To maximize losses
C. To minimize losses
D. To minimize the Q

C   Higher resistance means lower efficiency. In other words, as the resistance goes up so do the ohmic and system losses. You want to use a loading coil with a high ratio of reactance to resistance, because that minimizes losses.

E9D05   What is a disadvantage of using a multiband trapped antenna?
  A. It might radiate harmonics
  B. It radiates the harmonics and fundamental equally well
  C. It is too sharply directional at lower frequencies
  D. It must be neutralized

A   Because the trap antenna is a multiband antenna, it can do a good job of radiating harmonics.

E9D06   What happens to the bandwidth of an antenna as it is shortened through the use of loading coils?
  A. It is increased
  B. It is decreased
  C. No change occurs
  D. It becomes flat

B   The bandwidth will decrease if loading coils are used to shorten an antenna.

E9D07   What is an advantage of using top loading in a shortened HF vertical antenna?
  A. Lower Q
  B. Greater structural strength
  C. Higher losses
  D. Improved radiation efficiency

D   Top loading is a technique that can reduce loading-coil losses. The method requires that a capacitive hat be added above the coil. A capacitive hat is so-named because it consists of wire spokes or rings that have a hat-like appearance. The added capacitance at the top of the antenna allows a smaller value of loading inductance. This results in lower system losses, thus improving antenna radiation efficiency.

E9D08   What happens as the Q of an antenna increases?
  A. SWR bandwidth increases
  B. SWR bandwidth decreases
  C. Gain is reduced
  D. More common-mode current is present on the feed line

B   Q and bandwidth are inversely related. Higher Q means smaller bandwidth.

E9D09 What is the function of a loading coil used as part of an HF mobile antenna?

A. To increase the SWR bandwidth
B. To lower the losses
C. To lower the Q
D. To cancel capacitive reactance

D An HF mobile antenna is usually less than 1/4 wavelength long, which means the feed point has capacitive reactance. The loading coil is inductive and is used to tune out the capacitive reactance so that the antenna is resonant.

E9D10 What happens to feed point impedance at the base of a fixed length HF mobile antenna as the frequency of operation is lowered?

A. The radiation resistance decreases and the capacitive reactance decreases
B. The radiation resistance decreases and the capacitive reactance increases
C. The radiation resistance increases and the capacitive reactance decreases
D. The radiation resistance increases and the capacitive reactance increases

B Lowering the frequency of operation has the same effect as shortening the antenna. (The antenna becomes shorter electrically.) That means that the radiation resistance decreases and the capacitive reactance increases.

E9D11 Which of the following types of conductors would be best for minimizing losses in a station's RF ground system?

A. A resistive wire, such as a spark plug wire
B. A wide flat copper strap
C. A cable with six or seven 18 gauge conductors in parallel
D. A single 12 or 10 gauge stainless steel wire

B Ground conductors should have both low ohmic resistance and low inductance to minimize impedance at high frequencies. The type of conductor that best achieves those goals is a thin, flat copper strip or a large round copper wire.

E9D12 Which of the following would provide the best RF ground for your station?

A. A 50 ohm resistor connected to ground
B. An electrically short connection to a metal water pipe
C. An electrically short connection to 3 or 4 interconnected ground rods driven into the Earth
D. An electrically short connection to 3 or 4 interconnected ground rods via a series RF choke

C   A safety ground may consist of a single ground rod driven deep into the ground, but at RF, it is more important to distribute the RF current widely in the ground near the surface. For that reason, several shallow ground rods are often used, connected together as a single ground.

E9D13 What usually occurs if a Yagi antenna is designed solely for maximum forward gain?

A. The front-to-back ratio increases
B. The front-to-back ratio decreases
C. The frequency response is widened over the whole frequency band
D. The SWR is reduced

B   You can design a Yagi antenna for maximum forward gain, but the pattern ratios such as front-to-back and front-to-side ratio usually decrease.

E9E   Matching: matching antennas to feed lines; phasing lines; power dividers

E9E01 What system matches a higher impedance transmission line to a lower impedance antenna by connecting the line to the driven element in two places spaced a fraction of a wavelength each side of element center?

A. The gamma matching system
B. The delta matching system
C. The omega matching system
D. The stub matching system

B   The delta match network shown in Figure E9E01 matches a high-impedance transmission line to a lower impedance antenna by connecting the line to the driven element in two places spaced a fraction of a wavelength each side of the element center.

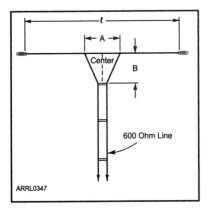

ARRL0347

Figure E9E01 — The delta matching system is used to match a high-impedance transmission line to a lower-impedance antenna. The feed line attaches to the driven element in two places, spaced a fraction of a wavelength on each side of the element center.

E9E02 What is the name of an antenna matching system that matches an unbalanced feed line to an antenna by feeding the driven element both at the center of the element and at a fraction of a wavelength to one side of center?

    A. The gamma match
    B. The delta match
    C. The epsilon match
    D. The stub match

A   The gamma matching system shown in Figure E9E02 matches an unbalanced feed line to an antenna by feeding the driven element both at the center of the element and at a fraction of a wavelength to one side of center.

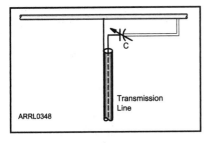

ARRL0348

Figure E9E02 — The gamma matching system is used to match an unbalanced feed line to an antenna. The feed line attaches to the center of the driven element and to a point that is a fraction of a wavelength to one side of center.

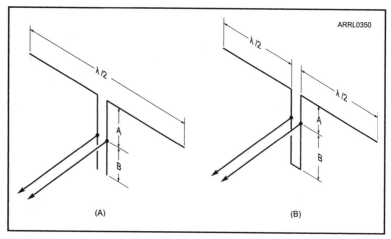

Figure E9E03 — The stub matching system uses a short perpendicular section of transmission line connected to the feed line near the antenna feed point.

E9E03   What is the name of the matching system that uses a section of transmission line connected in parallel with the feed line at or near the feed point?

    A. The gamma match
    B. The delta match
    C. The omega match
    D. The stub match

D   The stub matching system shown in Figure E9E03 uses a short perpendicular section of transmission line connected to the feed line near the antenna feed point.

E9E04   What is the purpose of the series capacitor in a gamma-type antenna matching network?

    A. To provide DC isolation between the feed line and the antenna
    B. To cancel the inductive reactance of the matching network
    C. To provide a rejection notch to prevent the radiation of harmonics
    D. To transform the antenna impedance to a higher value

B   See E9E02. The gamma match system's feed line section parallel to the antenna element has some inductive reactance at the feed point. The adjustable capacitor cancels the reactance and brings the antenna system to resonance.

E9E05 How must the driven element in a 3-element Yagi be tuned to use a hairpin matching system?

A. The driven element reactance must be capacitive
B. The driven element reactance must be inductive
C. The driven element resonance must be lower than the operating frequency
D. The driven element radiation resistance must be higher than the characteristic impedance of the transmission line

A See E9E06. The Yagi driven element must be tuned so it has capacitive reactance. The hairpin inductance then cancels the capacitive reactance as it transforms the feed point impedance to a higher value that matches that of the feed line.

E9E06 What is the equivalent lumped-constant network for a hairpin matching system of a 3-element Yagi?

A. Pi-network
B. Pi-L-network
C. A shunt inductor
D. A series capacitor

C The hairpin or beta match is an inductive network and the feed point is capacitive. Together they form the equivalent of an L-network. Part B of Figure E9E06 shows the lumped constant equivalent circuit, where R_A and C_A represent the antenna feed point impedance, and L represents the parallel inductance of the hairpin. Points X and Y represent the feed line connection. When the equivalent circuit is redrawn as shown at C, you can see that L and C_A form an L-network to match the feed line to the antenna resistance R_A.

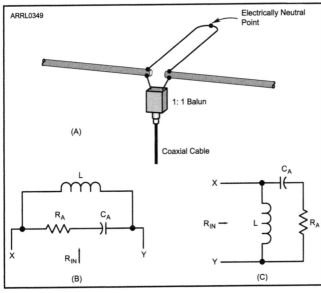

Figure E9E06 — The driven element of a Yagi antenna can be fed with a hairpin or beta matching system, as shown at A. Part B shows the lumped constant equivalent circuit.

E9E07   What term best describes the interactions at the load end of a mismatched transmission line?

A. Characteristic impedance
B. Reflection coefficient
C. Velocity factor
D. Dielectric constant

B   The reflection coefficient is the ratio of reflected voltage (or current) to the incident voltage (or current) at the same point on a line. The reflection coefficient is determined by the relationship between the feed line characteristic impedance and the actual load impedance. That makes the reflection coefficient a good parameter to describe the interactions at the load end of a mismatched transmission line.

E9E08   Which of the following measurements is characteristic of a mismatched transmission line?

A. An SWR less than 1:1
B. A reflection coefficient greater than 1
C. A dielectric constant greater than 1
D. An SWR greater than 1:1

D   If there is a mismatch, the SWR will be greater than 1:1. An SWR less than 1:1 is not possible. It is also not possible to have a dielectric constant or reflection coefficient greater than 1.

E9E09   Which of these matching systems is an effective method of connecting a 50 ohm coaxial cable feed line to a grounded tower so it can be used as a vertical antenna?

A. Double-bazooka match
B. Hairpin match
C. Gamma match
D. All of these choices are correct

C   A grounded tower can be thought of as one-half of a dipole driven element and a matching system used to match the base impedance of the tower to that of a feed line. The most convenient method of doing so is the gamma match.

E9E10 Which of these choices is an effective way to match an antenna with a 100 ohm feed point impedance to a 50 ohm coaxial cable feed line?

A. Connect a ¼ wavelength open stub of 300 ohm twin-lead in parallel with the coaxial feed line where it connects to the antenna
B. Insert a ½ wavelength piece of 300 ohm twin-lead in series between the antenna terminals and the 50-ohm feed cable
C. Insert a ¼ wavelength piece of 75 ohm coaxial cable transmission line in series between the antenna terminals and the 50 ohm feed cable
D. Connect a ½ wavelength shorted stub of 75 ohm cable in parallel with the 50-ohm cable where it attaches to the antenna

C A ¼-wavelength section of feed line with a characteristic impedance (Z_0) close to the geometric mean of the feed line impedance (Z_1) and antenna feed point impedance (Z_2), such that

$$Z_0 = \sqrt{Z_1 Z_2}$$

will match the two impedances. This is called a ¼-wave synchronous transformer.

E9E11 What is an effective way of matching a feed line to a VHF or UHF antenna when the impedances of both the antenna and feed line are unknown?

A. Use a 50-ohm 1:1 balun between the antenna and feed line
B. Use the universal stub matching technique
C. Connect a series-resonant LC network across the antenna feed terminals
D. Connect a parallel-resonant LC network across the antenna feed terminals

B See E9E03. The universal stub matching technique is illustrated in Figure E9E03. It is used at VHF and UHF where wavelengths are short enough for the technique to be practical.

E9E12 What is the primary purpose of a phasing line when used with an antenna having multiple driven elements?

A. It ensures that each driven element operates in concert with the others to create the desired antenna pattern

B. It prevents reflected power from traveling back down the feed line and causing harmonic radiation from the transmitter

C. It allows single-band antennas to operate on other bands

D. It makes sure the antenna has a low-angle radiation pattern

A Phased-array antennas depend on precise current and phase at the element feed points to create the desired patterns. Carefully constructed lengths of feed line are used so that the proper current and phasing relationships are created.

E9E13 What is a use for a Wilkinson divider?

A. It divides the operating frequency of a transmitter signal so it can be used on a lower frequency band

B. It is used to feed high-impedance antennas from a low-impedance source

C. It is used to divide power equally between two 50 ohm loads while maintaining 50 ohm input impedance

D. It is used to feed low-impedance loads from a high-impedance source

C The Wilkinson power divider is used to split the power from a single source into two or more equally divided portions for use in a phased array antenna system. It also helps isolate each output from each other and from the source.

E9F **Transmission lines: characteristics of open and shorted feed lines; ⅛ wavelength; ¼ wave length; ½ wavelength; feed lines: coax versus open-wire; velocity factor; electrical length; coaxial cable dielectrics; velocity factor**

E9F01 What is the velocity factor of a transmission line?

A. The ratio of the characteristic impedance of the line to the terminating impedance

B. The index of shielding for coaxial cable

C. The velocity of the wave in the transmission line multiplied by the velocity of light in a vacuum

D. The velocity of the wave in the transmission line divided by the velocity of light in a vacuum

D The velocity factor of a transmission line is the velocity of the wave in the line divided by the velocity of light in a vacuum.

E9F02   Which of the following determines the velocity factor of a transmission line?

A. The termination impedance
B. The line length
C. Dielectric materials used in the line
D. The center conductor resistivity

C   The presence of dielectric insulating materials reduces the velocity of an electromagnetic wave in a transmission line, since those waves travel more slowly in materials other than a vacuum.

E9F03   Why is the physical length of a coaxial cable transmission line shorter than its electrical length?

A. Skin effect is less pronounced in the coaxial cable
B. The characteristic impedance is higher in a parallel feed line
C. The surge impedance is higher in a parallel feed line
D. Electrical signals move more slowly in a coaxial cable than in air

D   The electrical length of a transmission line is measured in wavelengths at a given frequency. To calculate the physical length of line for a given electrical length, multiply the electrical (free space) length by the velocity factor.

E9F04   What is the typical velocity factor for a coaxial cable with solid polyethylene dielectric?

A. 2.70
B. 0.66
C. 0.30
D. 0.10

B   The typical velocity factor for a coaxial cable with polyethylene dielectric is 0.66. In other words, the speed of an electromagnetic wave in typical RG-8 coax is about two-thirds the speed of light in a vacuum.

E9F05   What is the approximate physical length of a solid polyethylene dielectric coaxial transmission line that is electrically one-quarter wavelength long at 14.1 MHz?

A. 20 meters
B. 2.3 meters
C. 3.5 meters
D. 0.2 meters

C   The electrical length of a transmission line is measured in wavelengths at a given frequency. To calculate the physical length of line for a given electrical length, multiply the electrical (free space) length by the velocity factor. The 14.1 MHz frequency corresponds to a free-space wavelength of 21.3 meter (300 divided by the frequency in MHz). Since we wish to have a ¼-wavelength line, the free-space length would be 5.3 meters. This is multiplied by the velocity factor to give 3.5 meters.

E9F06 What is the approximate physical length of an air-insulated, parallel conductor transmission line that is electrically one-half wavelength long at 14.10 MHz?

A. 15 meters
B. 20 meters
C. 10 meters
D. 71 meters

C The electrical length of a transmission line is measured in wavelengths at a given frequency. To calculate the physical length of line for a given electrical length, multiply the electrical (free space) length by the velocity factor. The free-space wavelength is 21.3 meters (300/14.1). We desire a ½-wavelength line, which would have a free-space length of 10.6 meters. This is multiplied by the velocity factor to give 10 meters.

E9F07 How does ladder line compare to small-diameter coaxial cable such as RG-58 at 50 MHz?

A. Lower loss
B. Higher SWR
C. Smaller reflection coefficient
D. Lower velocity factor

A Ladder line typically has less than one tenth the loss of RG-58 coaxial cable.

E9F08 What is the term for the ratio of the actual speed at which a signal travels through a transmission line to the speed of light in a vacuum?

A. Velocity factor
B. Characteristic impedance
C. Surge impedance
D. Standing wave ratio

A The ratio of the actual velocity at which a signal travels through a transmission line to the speed of light in a vacuum is called the velocity factor.

E9F09 What is the approximate physical length of a solid polyethylene dielectric coaxial transmission line that is electrically one-quarter wavelength long at 7.2 MHz?

A. 10 meters
B. 6.9 meters
C. 24 meters
D. 50 meters

B The free-space wavelength is 41.6 meters (300 divided by 7.2). You want a ¼-wavelength line, which would have a free-space length of 10.4 meters. Multiply this by the velocity factor (0.66) to give 6.9 meters.

Table E9-1
Properties of Open and Shorted Feed Line Sections

Length	Termination	Impedance
1/8 wavelength	Shorted	Inductive
1/8 wavelength	Open	Capacitive
1/4 wavelength	Shorted	Very high impedance
1/4 wavelength	Open	Very low impedance
1/2 wavelength	Shorted	Very low impedance
1/2 wavelength	Open	Very high impedance

E9F10 What impedance does a ⅛ wavelength transmission line present to a generator when the line is shorted at the far end?

  A. A capacitive reactance
  B. The same as the characteristic impedance of the line
  C. An inductive reactance
  D. The same as the input impedance to the final generator stage

C   This begins a series of questions about various line lengths and terminations. You'll find the answers in the table, which shows input impedance to various length line sections that are terminated in a short or an open circuit. For example, from Table E9-1, you see that a ⅛-wavelength transmission line that is shorted at the far end exhibits an inductive reactance at its input.

E9F11   What impedance does a ⅛ wavelength transmission line present to a generator when the line is open at the far end?

  A. The same as the characteristic impedance of the line
  B. An inductive reactance
  C. A capacitive reactance
  D. The same as the input impedance of the final generator stage

C   From Table E9-1, the open-ended ⅛ wavelength transmission line looks capacitive to the generator.

E9F12   What impedance does a ¼ wavelength transmission line present to a generator when the line is open at the far end?

  A. The same as the characteristic impedance of the line
  B. The same as the input impedance to the generator
  C. Very high impedance
  D. Very low impedance

D   From Table E9-1, the open-ended ¼-wavelength transmission line has a very low impedance at its input.

E9F13   What impedance does a ¼ wavelength transmission line present to a generator when the line is shorted at the far end?

    A. Very high impedance
    B. Very low impedance
    C. The same as the characteristic impedance of the transmission line
    D. The same as the generator output impedance

A   From Table E9-1, the shorted ¼ wavelength transmission line presents a very high impedance to the generator.

E9F14   What impedance does a ½ wavelength transmission line present to a generator when the line is shorted at the far end?

    A. Very high impedance
    B. Very low impedance
    C. The same as the characteristic impedance of the line
    D. The same as the output impedance of the generator

B   From Table E9-1, the shorted ½ wavelength transmission line presents a very low impedance at its input. It is useful to remember that impedance repeats every ½ wavelength along a transmission line.

E9F15   What impedance does a ½ wavelength transmission line present to a generator when the line is open at the far end?

    A. Very high impedance
    B. Very low impedance
    C. The same as the characteristic impedance of the line
    D. The same as the output impedance of the generator

A   From Table E9-1, the open 1/2 wavelength transmission line presents a very high impedance to the generator.

E9F16   Which of the following is a significant difference between foam dielectric coaxial cable and solid dielectric cable, assuming all other parameters are the same?

    A. Foam dielectric has lower safe operating voltage limits
    B. Foam dielectric has lower loss per unit of length
    C. Foam dielectric has higher velocity factor
    D. All of these choices are correct

D   The foam dielectric replaces some of the polyethylene with air, which has lower losses and increases the velocity factor. The tradeoff is that air is not quite as good an insulator as polyethylene and so the safe operating voltage limit is lower.

E9G   The Smith chart

E9G01   Which of the following can be calculated using a Smith chart?

A. Impedance along transmission lines
B. Radiation resistance
C. Antenna radiation pattern
D. Radio propagation

A   The Smith chart was developed for the purpose of graphically calculating impedance along a transmission line. It is unique in that ability.

E9G02   What type of coordinate system is used in a Smith chart?

A. Voltage circles and current arcs
B. Resistance circles and reactance arcs
C. Voltage lines and current chords
D. Resistance lines and reactance chords

B   The mnemonic for this question is R & R: resistance and reactance, circles and arcs. In this Smith chart impedances are plotted on a set of circles and arcs. Constant-resistance circles are made up of impedance values that all have the same resistance values. Constant-reactance arcs are the impedance points with the same values of reactance.

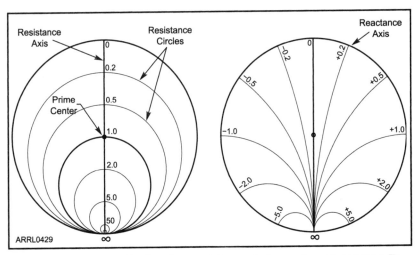

Figure E9G02 — A Smith chart has resistance circles and reactance arcs. To help you visualize them, they are shown separately in this drawing.

E9G03 Which of the following is often determined using a Smith chart?

A. Beam headings and radiation patterns
B. Satellite azimuth and elevation bearings
C. Impedance and SWR values in transmission lines
D. Trigonometric functions

C A Smith chart is not only used to plot impedances, it can also be used to calculate changes in resistance and reactance along a length of transmission line.

E9G04 What are the two families of circles and arcs that make up a Smith chart?

A. Resistance and voltage
B. Reactance and voltage
C. Resistance and reactance
D. Voltage and impedance

C Since the Smith chart graphs impedances, the correct answer is resistance and reactance.

E9G05 What type of chart is shown in Figure E9-3?

A. Smith chart
B. Free space radiation directivity chart
C. Elevation angle radiation pattern chart
D. Azimuth angle radiation pattern chart

A The chart in Figure E9-3 is a Smith chart.

E9G06 On the Smith chart shown in Figure E9-3, what is the name for the large outer circle on which the reactance arcs terminate?

A. Prime axis
B. Reactance axis
C. Impedance axis
D. Polar axis

B The reactance axis circle corresponds to the vertical reactance axis for rectangular coordinates.

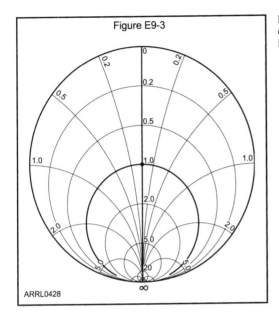

Figure E9-3

Figure E9-3 — Use this chart for E9G05 through E9G07.

ARRL0428

E9G07 On the Smith chart shown in Figure E9-3, what is the only straight line shown?

A. The reactance axis
B. The current axis
C. The voltage axis
D. The resistance axis

D The resistance axis corresponds to the horizontal resistance axis for rectangular coordinates.

E9G08 What is the process of normalization with regard to a Smith chart?

A. Reassigning resistance values with regard to the reactance axis
B. Reassigning reactance values with regard to the resistance axis
C. Reassigning impedance values with regard to the prime center
D. Reassigning prime center with regard to the reactance axis

C The process of assigning resistance values with regard to the value at prime center is called normalizing. To normalize values for a 50-ohm system, divide the resistance by 50. To convert from the chart values back to actual values, multiply by 50. Normalization permits the Smith chart to be used for any impedance value.

E9G09 What third family of circles is often added to a Smith chart during the process of solving problems?

A. Standing wave ratio circles
B. Antenna-length circles
C. Coaxial-length circles
D. Radiation-pattern circles

A Smith chart plots can be used to give a measure of impedance mismatch. All of the points on the chart that result in the same SWR form a circle centered on the prime center.

E9G10 What do the arcs on a Smith chart represent?

A. Frequency
B. SWR
C. Points with constant resistance
D. Points with constant reactance

D The reactance arcs correspond to horizontal lines in rectangular coordinates that have constant reactance.

E9G11 How are the wavelength scales on a Smith chart calibrated?

A. In fractions of transmission line electrical frequency
B. In fractions of transmission line electrical wavelength
C. In fractions of antenna electrical wavelength
D. In fractions of antenna electrical frequency

B The Smith chart was developed for the purpose of graphically calculating impedance along a transmission line. For that reason, the scale is calibrated in terms of wavelength in a transmission line.

E9H   Receiving antennas: radio direction finding antennas; Beverage antennas; specialized receiving antennas; longwire receiving antennas

E9H01 When constructing a Beverage antenna, which of the following factors should be included in the design to achieve good performance at the desired frequency?

A. Its overall length must not exceed 1/4 wavelength
B. It must be mounted more than 1 wavelength above ground
C. It should be configured as a four-sided loop
D. It should be one or more wavelengths long

D A simple Beverage antenna with terminating resistor and matching transformer for the feed line to the receiver. The Beverage antenna (invented by H.H. Beverage in 1922) works best when it is greater than one wavelength in length.

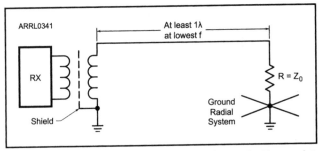

E9H01 — A simple Beverage antenna with terminating resistor and matching transformer for the feed line to the receiver.

E9H02 Which is generally true for low band (160 meter and 80 meter) receiving antennas?

A. Atmospheric noise is so high that gain over a dipole is not important
B. They must be erected at least 1/2 wavelength above the ground to attain good directivity
C. Low loss coax transmission line is essential for good performance
D. All of these choices are correct

A The reason for using special receiving antennas at these low frequencies is for their directivity to reject noise in some directions, increasing signal-to-noise ratio (SNR).

E9H03 DELETED February 1, 2016

E9H04 What is an advantage of using a shielded loop antenna for direction finding?

A. It automatically cancels ignition noise pickup in mobile installations
B. It is electrostatically balanced against ground, giving better nulls
C. It eliminates tracking errors caused by strong out-of-band signals
D. It allows stations to communicate without giving away their position

B A properly constructed shielded loop antenna is balanced against ground and so does not pick up stray signals that cause its pattern nulls to be shallower and less symmetrical. Both of those qualities are important in direction finding because they increase the precision and accuracy of the bearings to the signal source.

E9H05 What is the main drawback of a wire-loop antenna for direction finding?

A. It has a bidirectional pattern
B. It is non-rotatable
C. It receives equally well in all directions
D. It is practical for use only on VHF bands

A A wire-loop antenna has a bidirectional pattern. That means that it will not unambiguously identify the direction from which a signal is arriving.

E9H06 What is the triangulation method of direction finding?

A. The geometric angle of sky waves from the source are used to determine its position
B. A fixed receiving station plots three headings to the signal source
C. Antenna headings from several different receiving locations are used to locate the signal source
D. A fixed receiving station uses three different antennas to plot the location of the signal source

C To perform triangulation, combine azimuth measurements from several diverse physical locations to determine the transmitter location. These azimuths or bearings are typically drawn on a map, and where they cross is the location of the transmitter.

E9H07 Why is it advisable to use an RF attenuator on a receiver being used for direction finding?

A. It narrows the bandwidth of the received signal to improve signal to noise ratio
B. It compensates for the effects of an isotropic antenna, thereby improving directivity
C. It reduces loss of received signals caused by antenna pattern nulls, thereby increasing sensitivity
D. It prevents receiver overload which could make it difficult to determine peaks or nulls

D A direction-finding receiver will need to receive signals with a large power range. Since the receiver may have a limited reception signal range, an attenuator makes a convenient helper in extending the dynamic range of the receiver by attenuating signals from a nearby transmitter.

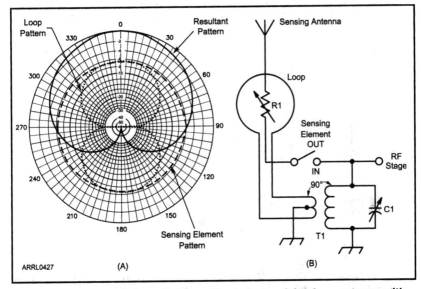

Figure E9H08 — At (A) the radiation pattern of a receiving loop antenna with sensing element. At (B) is a circuit for combining the signals from the two elements. C1 is adjusted for resonance at the operating frequency with T1.

E9H08 What is the function of a sense antenna?

A. It modifies the pattern of a DF antenna array to provide a null in one direction
B. It increases the sensitivity of a DF antenna array
C. It allows DF antennas to receive signals at different vertical angles
D. It provides diversity reception that cancels multipath signals

A A sense antenna (sometimes called a sensing element) is a vertical antenna that is added to a loop antenna as in the figure, and together they produce a cardioid reception pattern. This pattern is useful for direction finding.

E9H09 Which of the following describes the construction of a receiving loop antenna?

A. A large circularly polarized antenna
B. A small coil of wire tightly wound around a toroidal ferrite core
C. One or more turns of wire wound in the shape of a large open coil
D. A vertical antenna coupled to a feed line through an inductive loop of wire

C A receiving loop antenna consists of one or more turns of wire wound in the shape of a large open coil. Receiving loops are very short in terms of wavelengths while transmitting loops are generally approximately one wavelength or more in circumference.

E9H10 How can the output voltage of a multiple turn receiving loop antenna be increased?

A. By reducing the permeability of the loop shield
B. By increasing the number of wire turns in the loop and reducing the area of the loop structure
C. By winding adjacent turns in opposing directions
D. By increasing either the number of wire turns in the loop or the area of the loop structure or both

D   The strength of the signal coming from a loop antenna is proportional to the cross-sectional area of the antenna and the number of turns — the voltage increases as either of these parameters increase.

E9H11 What characteristic of a cardioid pattern antenna is useful for direction finding?

A. A very sharp peak
B. A very sharp single null
C. Broad band response
D. High-radiation angle

B   An antenna with either a sharp peak or a sharp null is good for direction finding because the peak or null can be used to point to the RF source. In the cardioid pattern, there is a deep, narrow null in one direction.

Subelement E0

Safety

There will be one question on your Extra class examination from the Safety subelement. This question will be taken from the group of questions labeled E0A.

E0A Safety: amateur radio safety practices; RF radiation hazards; hazardous materials; grounding

E0A01 What is the primary function of an external earth connection or ground rod?

A. Reduce received noise
B. Lightning protection
C. Reduce RF current flow between pieces of equipment
D. Reduce RFI to telephones and home entertainment systems

B The earth connection is primarily to establish earth potential during lightning strikes. All earth connections in a home and amateur station must be bonded together.

E0A02 When evaluating RF exposure levels from your station at a neighbor's home, what must you do?

A. Make sure signals from your station are less than the controlled MPE limits
B. Make sure signals from your station are less than the uncontrolled MPE limits
C. You need only evaluate exposure levels on your own property
D. Advise your neighbors of the results of your tests

B Because you do not control when your neighbor may be at home, the exposure is considered uncontrolled. When performing the required evaluation of your station, the uncontrolled MPE (maximum permissible exposure) limits are the ones you should use.

E0A03 Which of the following would be a practical way to estimate whether the RF fields produced by an amateur radio station are within permissible MPE limits?

A. Use a calibrated antenna analyzer
B. Use a hand calculator plus Smith chart equations to calculate the fields
C. Use an antenna modeling program to calculate field strength at accessible locations
D. All of the choices are correct

C   It is easiest by far to use the approved modeling technique to determine if RF exposure from your station is close to or exceeds the MPE limits in accessible locations. Should the modeling technique indicate excessive exposure, following up with on-site measurements could be warranted.

E0A04 When evaluating a site with multiple transmitters operating at the same time, the operators and licensees of which transmitters are responsible for mitigating over-exposure situations?

A. Only the most powerful transmitter
B. Only commercial transmitters
C. Each transmitter that produces 5 percent or more of its MPE limit at accessible locations
D. Each transmitter operating with a duty-cycle greater than 50 percent

C   In multitransmitter sites, such as hilltop facilities where there may be commercial facilities along with amateur repeaters, precisely determining responsibility for RF levels is impractical. Therefore, the rule was written so that responsibility is shared between the operators of all significant RF sources.

E0A05 What is one of the potential hazards of using microwaves in the amateur radio bands?

A. Microwaves are ionizing radiation
B. The high gain antennas commonly used can result in high exposure levels
C. Microwaves often travel long distances by ionospheric reflection
D. The extremely high frequency energy can damage the joints of antenna structures

B   Because the wavelength of microwaves is a few centimeters or less, it is relatively straightforward to construct antennas with gain far in excess of that on longer wavelength bands. This can cause very high field strengths in the main lobe of high-gain antennas such as dishes. Take special care when operating such antennas near people, as during portable operation.

E0A06 Why are there separate electric (E) and magnetic (H) field MPE limits?

A. The body reacts to electromagnetic radiation from both the E and H fields
B. Ground reflections and scattering make the field impedance vary with location
C. E field and H field radiation intensity peaks can occur at different locations
D. All of these choices are correct

D In the far field of antennas, exposure from the E and H fields can be combined into a composite power density. Near the antenna or near reflecting surfaces, the E and H field intensity can vary significantly from the far field values. The ratio of E to H field intensity — the media impedance — can also be altered by scattering and reflection. In such cases, it may be prudent to evaluate exposure based on both the E and H field values.

E0A07 How may dangerous levels of carbon monoxide from an emergency generator be detected?

A. By the odor
B. Only with a carbon monoxide detector
C. Any ordinary smoke detector can be used
D. By the yellowish appearance of the gas

B Since carbon monoxide (CO) is odorless and colorless, the only way to detect it is with a carbon monoxide detector. Similar to a smoke detector in appearance, CO detectors sound an alarm when unsafe concentrations of CO are present.

E0A08 What does SAR measure?

A. Synthetic Aperture Ratio of the human body
B. Signal Amplification Rating
C. The rate at which RF energy is absorbed by the body
D. The rate of RF energy reflected from stationary terrain

C Specific Absorption Rate (SAR) measures the rate at which energy from an electromagnetic field is absorbed by the human body. SAR changes with frequency and the shape of the body part.

E0A09 Which insulating material commonly used as a thermal conductor for some types of electronic devices is extremely toxic if broken or crushed and the particles are accidentally inhaled?

A. Mica
B. Zinc oxide
C. Beryllium oxide
D. Uranium hexaflouride

C Beryllium oxide (BeO or beryllia) is a white ceramic insulator that has the unusual property of also being an excellent thermal conductor. Most insulators are poor conductors of heat, so BeO is used in electronic devices as an insulating layer between the semiconductor itself and the metal case and in some vacuum tubes. BeO is not commonly encountered, but if there is any possibility of exposure, use protective gloves and take care not to touch or breath in any BeO dust. Check the manufacturer's data sheet to see if BeO is used in specific devices.

E0A10 What toxic material may be present in some electronic components such as high voltage capacitors and transformers?

A. Polychlorinated biphenyls
B. Polyethylene
C. Polytetrafluoroethylene
D. Polymorphic silicon

A Polychlorinated biphenyls (PCBs) were once commonly added to insulating oils to improve their stability and insulating qualities. It was not discovered until later that they are carcinogenic (increase the probability of developing cancer) and so commercial production and use halted. Amateurs may encounter PCBs in the insulating oil of old oil-filled capacitors or dummy loads. To dispose of these components safely, consult your local electric utility for information.

E0A11 Which of the following injuries can result from using high-power UHF or microwave transmitters?

A. Hearing loss caused by high voltage corona discharge
B. Blood clotting from the intense magnetic field
C. Localized heating of the body from RF exposure in excess of the MPE limits
D. Ingestion of ozone gas from the cooling system

C Commercial equipment using klystrons or magnetrons (including microwave ovens) is carefully shielded to avoid microwave leakage. Homemade or converted equipment may not have the same extensive shielding and so should be operated with caution. The primary hazard is heating as the body absorbs the microwave energy.

Notes

Notes

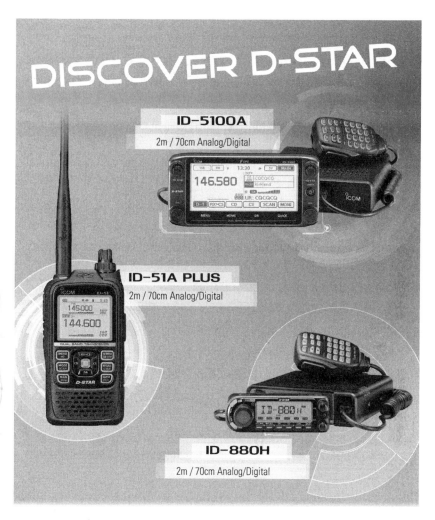

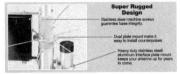